The Moon

by Thomas Gwyn Elger

PREFACE

This book and the accompanying map is chiefly intended for the use of lunar observers, but it is hoped it may be acceptable to many who, though they cannot strictly be thus described, take a general interest in astronomy.

The increasing number of those who possess astronomical telescopes, and devote more or less of their leisure in following some particular line of research, is shown by the great success in recent years of societies, such as the British Astronomical Association with its several branches, the Astronomical Society of the Pacific, and similar institutions in various parts of the world. These societies are not only doing much in popularising the sublimest of the sciences, but are the means of developing and organising the capabilities of their members by discouraging aimless and desultory observations, and by pointing out how individual effort may be utilised and made of permanent value in almost every department of astronomy.

The work of the astronomer, like that of the votary of almost every other science, is becoming every year more and more specialised; and among its manifold subdivisions, the study of the physical features of the moon is undoubtedly increasing in popularity and importance. To those who are pursuing such observations, it is believed that this book will be a useful companion to the telescope, and convenient for reference.

Great care has been taken in the preparation of the map, which, so far as the positions of the various objects represented are concerned, is based on the last edition of Beer and Madler's chart, and on the more recent and much larger and elaborate map of Schmidt; while as regards the shape and details of most of the formations, the author's drawings and a large number of photographs have been utilised. Even on so small a scale as eighteen inches to the moon's diameter, more detail might have been inserted, but this, at the expense of distinctness, would have detracted from the value of the map for handy reference in the usually dim light of the observatory, without adding to its utility in other ways. Every named formation is prominently shown; and most other features of interest, including the principal rill-systems, are represented, though, as regards these, no attempt is made to indicate all their manifold details and ramifications, which, to do effectually, would in very many instances require a map on a much larger scale than any that has yet appeared.

The insertion of meridian lines and parallels of latitude at every ten degrees, and the substitution of names for reference numbers, will add to the usefulness of the map.

With respect to the text, a large proportion of the objects in the Catalogue and in the Appendix have been observed and drawn by the author many times during the last thirty years, and described in _The Observatory_ and other publications. He has had, besides, the advantage of consulting excellent sketches by Mr W.H. MAW, F.R.A.S., Dr. SHELDON, F.R.A.S., Mr. A. MEE, F.R.A.S., Mr. G.P. HALLOWES, F.R.A.S., Dr. SMART, F.R.A.S., Mr. T. GORDON, F.R.A.S., Mr. G.T. DAVIS, Herr BRENNER, Herr KRIEGER, Mr. H. CORDER, and other members of the British Astronomical Association. Through the courtesy of Professor HOLDEN, Director of the Lick Observatory, and M. PRINZ, of the Royal Observatory of Brussels, many beautiful photographs and direct photographic enlargements have been available, as have also the exquisite heliogravures received by the author from Dr. L. WEINEK, Director of the Imperial Observatory of Prague, and the admirable examples of the photographic work of MM. PAUL and PROSPER HENRY of the Paris Observatory, which are occasionally published in Knowledge. The numerous representations of lunar objects which have appeared from time to time in that storehouse of astronomical information, The English Mechanic, and the invaluable notes in "Celestial Objects for Common Telescopes," and in various periodicals, by the late REV. PREBENDARY WEBB, to whom Selenography and Astronomy generally owe so much, have also been consulted.

As a rule, all the more prominent and important features are described, though very frequently interesting details are referred to which, from their minuteness, could not be shown in the map. The measurements (given in round numbers) are derived in most instances from NEISON'S (Nevill) "Moon," though occasionally those in the introduction to Schmidt's chart are adopted.

THOMAS GYWN ELGER. BEDFORD, 1895.

CONTENTS

INTRODUCTION MARIA, OR PLAINS, TERMED "SEAS" RIDGES RING-MOUNTAINS, CRATERS, &C. Walled Plains Mountain Rings Ring-Plains Craters Crater Cones Craterlets, Crater Pits MOUNTAIN RANGES, ISOLATED MOUNTAINS, &c. CLEFTS, OR RILLS FAULTS VALLEYS BRIGHT RAY-SYSTEMS THE MOON'S ALBEDO, SURFACE BRIGHTNESS, &c. TEMPERATURE OF THE MOON'S SURFACE LUNAR OBSERVATION PROGRESS OF SELENOGRAPHY, LUNAR PHOTOGRAPHY

CATALOGUE OF LUNAR FORMATIONS FIRST QUADRANT-- West Longitude 90 deg. to 60 deg. West Longitude 60 deg. to 40 deg. West Longitude 40 deg. to 20 deg. West Longitude 20 deg. to 0 deg. SECOND QUADRANT-- East Longitude 0 deg. to 20 deg. East Longitude 20 deg. to 40 deg. East Longitude 40 deg. to 60 deg. East Longitude 60 deg. to 90 deg. THIRD QUADRANT-- East Longitude 0 deg. to 20 deg. East Longitude 20 deg. to 40 deg. East Longitude 40 deg. to 60 deg. East Longitude 60 deg. to 90 deg. FOURTH QUADRANT-- West Longitude 90 deg. to 60 deg. West Longitude 60 deg. to 40 deg. West Longitude 40 deg. to 20 deg. West Longitude 20 deg. to 0 deg.

MAP OF THE MOON First Quadrant Second Quadrant Third Quadrant Fourth Quadrant

APPENDIX Description of Map List of the Maria, or Grey Plains, termed "Seas," &c. List of some of the most Prominent Mountain Ranges, Promontories, Isolated Mountains, and Remarkable Hills List of the Principal Ray-Systems, Light-Surrounded Craters, and Light Spots Position of the Lunar Terminator Lunar Elements Alphabetical List of Formations

INTRODUCTION

We know, both by tradition and published records, that from the earliest times the faint grey and light spots which diversify the face of our satellite excited the wonder and stimulated the curiosity of mankind, giving rise to suppositions more or less crude and erroneous as to their actual nature and significance. It is true that Anaxagoras, five centuries before our era, and probably other philosophers preceding him, --certainly Plutarch at a much later date--taught that these delicate markings and differences of tint, obvious to every one with normal vision, point to the existence of hills and valleys on her surface; the latter maintaining that the irregularities of outline presented by the "terminator," or line of demarcation between the illumined and unillumined portion of her spherical superficies, are due to mountains and their shadows; but more than fifteen centuries elapsed before the truth of this sagacious conjecture was unquestionably demonstrated. Selenography, as a branch of observational astronomy, dates from the spring of 1609, when Galileo directed his "optic tube" to the moon, and in the following year, in the Sidereus Nuncius, or "the Intelligencer of the Stars," gave to an astonished and incredulous world an account of the unsuspected marvels it revealed. In this remarkable little book we have the first attempt to represent the telescopic aspect of the moon's visible surface in the five rude woodcuts representing the curious features he perceived thereon, whose form and arrangement, he tells us, reminded him of the "ocelli" on the feathers of a peacock's tail,--a quaint but not altogether inappropriate simile to describe the appearance of groups of the larger ring-mountains partially illuminated by the sun, when seen in a small telescope.

The bright and dusky areas, so obvious to the unaided sight, were found by Galileo to be due to a very manifest difference in the character of the lunar surface, a large portion of the northern hemisphere, and no inconsiderable part of the south-eastern quadrant, being seen to consist of large grey monotonous tracts, often bordered by lofty mountains, while the remainder of the superficies was much more conspicuously brilliant, and, moreover, included by far the greater number of those curious ring- mountains and other extraordinary features whose remarkable aspect and peculiar arrangement first attracted his attention. Struck by the analogy which these contrasted regions present to the land and water surfaces of our globe, he suspected that the former are represented on the moon by the brighter and more rugged, and the latter by the smoother and more level areas; a view, however, which Kepler more distinctly formulated in the dictum, "Do maculas esse Maria, do lucidas esse terras." Besides making a rude lunar chart, he estimated the heights of some of the ring- mountains by measuring the distance from the terminator of their bright summit peaks, when they were either coming into or passing out of sunlight; and though his method was incapable of accuracy, and his results consequently untrustworthy, it served to demonstrate the immense altitude of these circumvallations, and to show how greatly they exceed any mountains on the earth if the relative dimensions of the two globes are taken into consideration.

Before the close of the century when selenography first became possible, Hevel of Dantzig, Scheiner, Langrenus (cosmographer to the King of Spain), Riccioli, the Jesuit astronomer of Bologna, and Dominic Cassini, the celebrated French astronomer, greatly extended the knowledge of the moon's surface, and published drawings of various phases, and charts, which, though very rude and incomplete, were a clear advance upon what Galileo, with his inferior optical means, had been able to accomplish. Langrenus, and after him Hevel, gave distinctive names to the various formations, mainly derived from terrestrial physical features, for which Riccioli subsequently substituted those of philosophers, mathematicians, and other celebrities; and Cassini determined by actual measurement the relative position of many of the principal objects on the disc, thus laying the foundation of an accurate system of lunar topography; while the labours of T. Mayer and Schroter in the last century, and of Lohrmann, Madler, Neison (Nevill), Schmidt, and other observers in the present, have been mainly devoted to the study of the minuter detail of the moon and its physical characteristics.

As was manifest to the earliest telescopic observers, its visible surface is clearly divisible into strongly contrasted areas, differing both in colour and structural character. Somewhat less than half of what we see of it consists of comparatively level dark tracts, some of them very many thousands of square miles in extent, the monotony of whose dusky superficies is often unrelieved for great distances by any prominent object; while the remainder, everywhere manifestly brighter, is not only more rugged and uneven, but is covered to a much greater extent with numbers of quasi-circular formations, differing widely in size, classed as walled-plains, ring-plains, craters, craterlets, crater-cones, &c. (the latter bearing a great outward resemblance to some terrestrial volcanoes), and mountain ranges of vast proportions, isolated hills, and other features.

Though nothing resembling sheets of water, either of small or large extent, have ever been detected on the surface, the superficial resemblance, in small telescopes, of the large grey tracts to the appearance which we may suppose our terrestrial lakes and oceans would present to an observer on the moon, naturally induced the early selenographers to term them Maria, or "seas"--a convenient name, which is still maintained, without, however, implying that these areas, as we now see them, are, or ever were, covered with water. Some, however, regard them as old sea-beds, from which every trace of fluid, owing to some unknown cause, has vanished, and that the folds and wrinkles, the ridges, swellings, and other peculiarities of structure observed upon them, represent some of the results of alluvial action. It is, of course, possible, and even probable, that at a remote epoch in the evolution of our satellite these lower regions were occupied by water, but that their surface, as it now appears, is actually this old sea-bottom, seems to be less likely than that it represents the consolidated crust of some semi- fluid or viscous material (possibly of a basaltic type) which has welled forth from orifices or rents communicating with the interior, and overspread and partially filled up these immense hollows, more or less overwhelming and destroying many formations which stood upon them before this catastrophe took place. Though this, like many other speculations of a similar character relating to lunar "geology," must remain, at least for the present, as a mere hypothesis; indications of this partial destruction by some agency or other is almost everywhere apparent in those formations which border the so-called seas, as, for example, Fracastorius in the Mare Nectaris; Le Monnier in the Mare Serenitatis; Pitatus and Hesiodus, on the south side of the Mare Nubium; Doppelmayer in the Mare Humorum, and in many other situations; while no observer can fail to notice innumerable instances of more or less complete obliteration and ruin among objects within these areas, in the form of obscure rings (mere scars on the surface), dusky craters, circular arrangements of isolated hills, reminding one of the monoliths of a Druidical temple; all of which we are justified in concluding were at one time formations of a normal type. It has been held by some selenologists --and Schmidt appears to be of the number,--that, seeing the comparative scarcity of large ring-plains and other massive formations on the Maria, these grey plains represent, as it were, a picture of the primitive surface of the moon before it was disturbed by the operations of interior forces; but this view affords no explanation of the undoubted existence of the relics of an earlier lunar world beneath their smooth superficies.

MARIA.--Leaving, however, these considerations for a more particular description of the Maria, it is clearly impossible, in referring to their level relatively to the higher and brighter land surface of the moon, to appeal to any hypsometrical standard. All that is known in this respect is, that they are invariably lower than the latter, and that some sink to a greater depth than others, or, in other words, that they do not all form a part of the same sphere. Though they are more or less of a greyish-slaty hue--some of them approximating very closely to that of the pigment known as "Payne's grey"--the tone, of course, depends upon the angle at which the solar rays impinge on that particular portion of the surface under observation. Speaking generally, they are, as would follow from optical considerations, conspicuously darker when viewed near the terminator, or when the sun is either rising or setting upon them, than under a more vertical angle of illumination. But even when it is possible to compare their colour by eye-estimation under similar solar altitudes, it is found that not only are some of the Maria, as a whole, notably darker than others, but nearly all of them exhibit local inequalities of hue, which, under good atmospheric and instrumental conditions, are especially remarkable. Under such circumstances I have frequently seen the surface, in many places covered with minute glittering points of light, shining with a silvery lustre, intermingled with darker spots and a network of streaks far too delicate and ethereal to represent in a drawing. In addition to these contrasts and differences in the sombre tone of these extended plains, many observers have remarked traces of a yellow or green tint on the surface of some of them. For example, the Mare Imbrium and the Mare Frigoris appear under certain conditions to be of a dirty yellow-green hue, the central parts of the Mare Humorum dusky green, and part of the Mare Serenitatis and the Mare Crisium light green, while the Palus Somnii has been noted a golden-brown yellow. To these may be added the district round Taruntius in the Mare Foecunditatis, and portions of other regions referred to in the catalogue, where I have remarked a very decided sepia colour under a low sun. It has been attempted to account for these phenomena by supposing the existence of some kind of vegetation; but as this involves the presence of an atmosphere, the idea hardly finds favour at the present time, though perhaps the possibility of plant growth in the low-lying districts, where a gaseous medium may prevail, is not altogether so chimerical a notion as to be unworthy of consideration. Nasmyth and others suggest that these tints may be due to broad expanses of coloured volcanic material, an hypothesis which, if we believe the Maria to be overspread with such matter, and knowing how it varies in colour in terrestrial volcanic regions, is more probable than the first. Anyway, whether we consider these appearances to be objective, or, after all, only due to purely physiological causes, they undoubtedly merit closer study and investigation than they have hitherto received.

There are twenty-three of these dusky areas which have received distinctive names; seventeen of them are wholly, or in great part, confined to the northern, and to the south-eastern quarter of the southern hemisphere--the south-western quadrant being to a great extent devoid of them. By far the largest is the vast Oceanus Procellarum, extending from a high northern latitude to beyond latitude 10 deg. in the south-eastern quadrant, and, according to Schmidt, with its bays and inflections, occupying an area of nearly two million square miles, or more than that of all the remaining Maria put together. Next in order of size come the Mare Nubium, of about one-fifth the superficies, covering a large portion of the south-eastern quadrant, and extending considerably north of the equator, and the Mare Imbrium, wholly confined to the northeastern quadrant, and including an area of about 340,000 square miles. These are by far the largest lunar "seas." The Mare Foecunditatis, in the western hemisphere, the greater part of it lying in the south- western quadrant, is scarcely half so big as the Mare Imbrium; while the Maria Serenitatis and Tranquilitatis, about equal in area (the former situated wholly north of the equator, and the latter only partially extending south of it), are still smaller. The arctic Mare Frigoris, some 100,000 square miles in extent, is the only remaining large sea,--the rest, such as the Mare Vaporum, the Sinus Medii, the Mare Crisium, the Mare Humorum, and the Mare Humboldtianum, are of comparatively small dimensions, the Mare Crisium not greatly exceeding 70,000 square miles, the Mare Humorum (about the size of England) 50,000 square miles, while the Mare Humboldtianum, according to Schmidt, includes only about 42,000 square miles, an area which is approached by some formations not classed with the Maria. This distinction, speaking generally, prevails among the Maria,--those of larger size, such as the Oceanus Procellarum, the Mare Nubium, and the Mare Foecunditatis, are less definitely enclosed, and, like terrestrial oceans, communicate with one another; while their borders, or, if the term may be allowed, their coast-line, is often comparatively low and ill-defined, exhibiting many inlets and irregularities in outline. Others, again, of considerable area, as, for example, the Mare Serenitatis and the Mare Imbrium, are bounded more or less completely by curved borders, consisting of towering mountain ranges, descending with a very steep escarpment to their surface: thus in form and other characteristics they resemble immense wall-surrounded plains. Among the best examples of enclosed Maria is the Mare Crisium, which is considered by Neison to be the deepest of all, and the Mare

Humboldtianum.

Though these great plains are described as level, this term must only be taken in a comparative sense. No one who observes them when their surface is thrown into relief by the oblique rays of the rising or setting sun can fail to remark many low bubble-shaped swellings with gently rounded outlines, shallow trough-like hollows, and, in the majority of them, long sinuous ridges, either running concentrically with their borders or traversing them from side to side. Though none of these features are of any great altitude or depth, some of the ridges are as much as 700 feet in height, and probably in many instances the other elevations often rise to 150 feet or more above the low-lying parts of the plains on which they stand. Hence we may say that the Maria are only level in the sense that many districts in the English Midland counties are level, and not that their surface is absolutely flat. The same may be said as to their apparent smoothness, which, as is evident when they are viewed close to the terminator, is an expression needing qualification, for under these conditions they often appear to be covered with wrinkles, flexures, and little asperities, which, to be visible at all, must be of considerable size. In fact, were it possible to examine them from a distance of a few miles, instead of from a standpoint which, under the most favourable circumstances, cannot be reckoned at less than 300, and this through an interposed aerial medium always more or less perturbed, they would probably be described as rugged and uneven, as some modern lava sheets.

RIDGES.--Among the Maria which exhibit the most remarkable arrangement of ridges is the Mare Humorum, in the south-eastern quadrant. Here, if it be observed under a rising sun, a number of these objects will be seen extending from the region north of the ring-mountain Vitello in long undulating lines, roughly concentric with the western border of the "sea," and gradually diminishing in altitude as they spread out, with many ramifications, to a distance of 200 miles or more towards the north. At this stage of illumination they are strikingly beautiful in a good telescope, reminding one of the ripple-marks left by the tide on a soft sandy beach. Like most other objects of their class, they are very evanescent, gradually disappearing as the sun rises higher in the lunar firmament, and ultimately leaving nothing to indicate their presence beyond here and there a ghostly streak or vein of a somewhat lighter hue than that of the neighbouring surface. The Mare Nectaris, again, in the south-western quadrant, presents some fine examples of concentric ridges, which are seen to the best advantage when the morning sun is rising on Rosse, a prominent crater north of Fracastorius. This "sea" is evidently concave in cross-section, the central portion being considerably lower than the margin, and these ridges appear to mark the successive stages of the change of level from the coast-line to the centre. They suggest the "caving in" of the surface, similar to that observed on a frozen pond or river, where the "cat's ice" at the edge, through the sinking of the water beneath, is rent and tilted to a greater or less degree. The Mare Serenitatis and the Mare Imbrium, in the northern hemisphere, are also remarkable for the number of these peculiar features. They are very plentifully distributed round the margin and in other parts of the former, which includes besides one of the longest and loftiest on the moon's visible surface--the great serpentine ridge, first drawn and described nearly a hundred years ago by the famous selenographer, Schroter of Lilienthal. Originating at a little crater under the north- east wall of great ring-plain Posidonius, it follows a winding course across the Mare toward the south, throwing out many minor branches, and ultimately dies out under a great rocky promontory--the Promontory Acherusia, at the western termination of the Haemus range. A comparatively low power serves to show the curious structural character of this immense ridge, which appears to consist of a number of corrugations and folds massed together, rising in places, according to Neison, to a height of 700 feet and more. The Mare Imbrium also affords an example of a ridge, which, though shorter, is nearly as prominent, in that which runs from the bright little ring-plain Piazzi Smyth towards the west side of Plato. The region round Timocharis and other quarters of the Mare are likewise traversed by very noteworthy features of a similar class. The Oceanus Procellarum also presents good instances of ridges in the marvellous ramifications round Encke, Kepler, and Marius, and in the region north of Aristarchus and Herodotus. Perhaps the most perfect examples of surface swellings are those in the Mare Tranquilitatis, a little east of the ring-plain Arago, where there are two nearly equal circular mounds, at least ten miles in diameter, resembling tumuli seen from above. Similar, but more irregular, objects of a like kind are very plentiful in many other quarters.

It is a suggestive peculiarity of many of the lunar ridges, both on the Maria and elsewhere, that they are very generally found in association with craters of every size. Illustrations of this fact occur almost everywhere. Frequently small craters are found on the summits of these elevations, but more often on their flanks and near their base. Where a ridge suddenly changes its direction, a crater of some prominence generally marks the point, often forming a node, or crossing-place of other ridges, which thus appear to radiate from it as a centre. Sometimes they intrude within the smaller ring-mountains, passing through gaps in their walls as, for example, in the cases of Madler, Lassell, &c. Various hypotheses have been advanced to account for them. The late Professor Phillips, the geologist, who devoted much attention to the telescopic examination of the physical features of the moon, compared the lunar ridges to long, low, undulating mounds, of somewhat doubtful origin, called "kames" in Scotland, and "eskers" in Ireland, where on the low central plain they are commonly found in the form of extended banks (mainly of gravel), with more or less steep sides, rising to heights of from 20 to 70 feet. They are sometimes only a few yards wide at the top, while in other places they spread out into large humps, having circular or oval cavities on their summits, 50 or 60 yards across, and as much as 40 feet deep. Like the lunar ridges, they throw out branches and exhibit many breaches of continuity. By some geologists they are supposed to represent old submarine banks formed by tidal currents, like harbour bars, and by others to be glacial deposits; in either case, to be either directly or indirectly due to alluvial action. Their outward resemblance to some of the ridges on the moon is unquestionable; and if we could believe that the Maria, as we now see them, are dried-up sea-beds, it might be concluded that these ridges had a similar origin; but their close connection with centres of volcanic disturbance, and the numbers of little craters on or near their track, point to the supposition that they consist rather of material exuded from long-extending fissures in the crust of the "seas," and in other surfaces where they are superimposed. This conjecture is rendered still more probable by the fact that we sometimes find the direction of clefts (which are undoubted surface cracks) prolonged in the form of long narrow ridges or of rows of little hillocks. We are, however, not bound to assume that all the manifold corrugations observed on the lunar plains are due to one and the same cause; indeed, it is clear that some are merely the outward indications of sudden drops in the surface, as in the case of the ridges round the western margin of the Mare Nectaris, and in other situations, where subsidence is manifested by features assuming the outward aspect of ordinary ridges, but which are in reality of a very different structural character.

The Maria, like almost every other part of the visible surface, abound in craters of a minute type, which are scattered here and there without any apparent law or ascertained principle of arrangement. Seeing how imperfect is our acquaintance with even the larger objects of this class, it is rash to insist on the antiquity or permanence of such diminutive objects, or to dogmatise about the cessation of lunar activity in connection with features where the volcanic history of our globe, if it is of any value as an analogue, teaches us it is most likely to prevail.

Most observers will agree with Schmidt, that observations and drawings of objects on the sombre depressed plains of the moon are easier and pleasanter to make than on the dazzling highlands, and that the lunar "sea" is to the working selenographer like an oasis in the desert to the traveller--a relief in this case, however, not to an exhausted body, but to a weary eye.

RING-MOUNTAINS, CRATERS, &C.--It is these objects, in their almost endless variety and bewildering number, which, more than any others, give to our satellite that marvellous appearance in the telescope which since the days of Galileo has never failed to evoke the astonishment of the beholder. However familiar we may be with the lunar surface, we can never gaze on these extraordinary formations, whether massed together apparently in inextricable confusion, or standing in isolated grandeur, like Copernicus, on the grey surface of the plains, without experiencing, in a scarcely diminished degree, the same sensation of wonder and admiration with which they were beheld for the first time. Although the attempt to bring all these bizarre forms under a rigid scheme of classification has not been wholly successful, their structural peculiarities, the hypsometrical relation between their interior and the surrounding district, their size, and the character of their circumvallation, the dimensions of their cavernous opening as compared with that of the more or less truncated conical mass of matter surrounding it, all afford a basis for grouping them under distinctive titles, that are not only convenient to the selenographer, but which undoubtedly represent, as a rule, actual diversities in their origin and physical character.

These distinguishing titles, as adopted by Schroter, Lohrmann, and Madler, and accepted by subsequent observers, are WALLED-PLAINS, MOUNTAIN RINGS, RING-PLAINS, CRATERS, CRATER-CONES, CRATERLETS, CRATER-PITS, DEPRESSIONS.

WALLED-PLAINS.--These formations, approximating more or less to the circular form, though frequently deviating considerably from it, are among the largest enclosures on the moon. They vary from upwards of 150 to 60 miles or under in diameter, and are often encircled by a complex rampart of considerable breadth, rising in some instances to a height of 12,000 feet or more above the enclosed plain. This rampart is rarely continuous, but is generally interrupted by gaps, crossed by transverse valleys and passes, and broken by more recent craters and depressions. As a rule, the area within the circumvallation (usually termed "the floor") is only slightly, if at all, lower than the region outside: it is very generally of a dusky hue, similar to that of the grey plains or Maria, and, like them, is usually variegated by the presence of hills, ridges, and craters, and is sometimes traversed by delicate furrows, termed clefts or rills.

Ptolemaeus, in the third quadrant, and not far removed from the centre of the disc, may be taken as a typical example of the class. Here we have a vast plain, 115 miles from side to side, encircled by a massive but much broken wall, which at one peak towers more than 9000 feet above a level floor, which includes details of a very remarkable character. The adjoining Alphonsus is another, but somewhat smaller, object of the same type, as are also Albategnius, and Arzachel; and Plato, in a high northern latitude, with its noble many-peaked rampart and its variable steel-grey interior. Grimaldi, near the eastern limb (perhaps the darkest area on the moon), Schickard, nearly as big, on the south- eastern limb, and Bailly, larger than either (still farther south in the same quadrant), although they approach some of the smaller "seas" in size, are placed in the same category. The conspicuous central mountain, so frequently associated with other types of ringed enclosures, is by no means invariably found within the walled-plains; though, as in the case of Petavius, Langrenus, Gassendi, and several other noteworthy examples, it is very prominently displayed. The progress of sunrise on all these objects affords a magnificent spectacle. Very often when the rays impinge on their apparently level floor at an angle of from 1 deg. to 2 deg., it is seen to be coarse, rough grained, and covered with minute elevations, although an hour or so afterwards it appears as smooth as glass.

Although it is a distinguishing characteristic that there is no great difference in level between the outside and the inside of a walled-plain, there are some very interesting exceptions to this rule, which are termed by Schmidt "Transitional forms." Among these he places some of the most colossal formations, such as Clavius, Maurolycus, Stofler, Janssen, and Longomontanus. The first, which may be taken as representative of the class (well known to observers as one of the grandest of lunar objects), has a deeply sunken floor, fringed with mountains rising some 12,000 feet above it, though they scarcely stand a fourth of this height above the plain on the west, which ascends with a very gentle gradient to the summit of the wall. Hence the contrast between the shadows of the peaks of the western wall on the floor at sunrise, and of the same peaks on the region west of the border at sunset is very marked. In Gassendi, Phocylides, and Wargentin we have similar notable departures from the normal type. The floor of the former on the north stands 2000 feet above the Mare Humorum. In Phocylides, probably through "faulting," one portion of the interior suddenly sinks to a considerable depth below the remainder; while the very abnormal Wargentin has such an elevated floor, that, when viewed under favourable conditions, it reminds one of a shallow oval tray or dish filled with fluid to the point of overflowing. These examples, very far from being exhaustive, will be sufficient to show that the walled-plains exhibit noteworthy differences in other respects than size, height of rampart, or included detail. Still another peculiarity, confined, it is believed, to a very few, may be mentioned, viz., convexity of floor, prominently displayed in Petavius, Mersenius, and Hevel.

MOUNTAIN RINGS.--These objects, usually encircled by a low and broken border, seldom more than a few hundred feet in height, are closely allied to the walled-plains. They are more frequently found on the Maria than elsewhere. In some cases the ring consists of isolated dark sections, with here and there a bright mass of rock interposed; in others, of low curvilinear ridges, forming a more or less complete circumvallation. They vary in size from 60 or 70 miles to 15 miles and less. The great ring north of Flamsteed, 60 miles across, is a notable example; another lies west of it on the north of Wichmann; while a third will be found south- east of Encke;--indeed, the Mare Procellarum abounds in objects of this type. The curious formation on the Mare Imbrium immediately south of Plato (called "Newton" by Schroter), may be placed in this category, as may also many of the low dusky rings of much smaller dimensions found in many quarters of the Maria. As has been stated elsewhere, these features have the appearance of having once been formations of a much more prominent and important character, which have suffered destruction, more or less complete, through being partially overwhelmed by the material of the "seas."

RING-PLAINS.--These are by far the most numerous of the ramparted enclosures of the moon, and though it is occasionally difficult to decide in which class, walled-plain or ring-plain, some objects should be placed, yet, as a rule, the difference between the structural character of the two is abundantly obvious. The ring-plains vary in diameter from sixty to less than ten miles, and are far more regular in outline than the walled-plains. Their ramparts, often very massive, are more continuous, and fall with a steep declivity to a floor almost always greatly depressed below the outside region. The inner slopes generally display subordinate heights, called terraces, arranged more or less concentrically, and often extending in successive stages nearly down to the interior foot of the wall. With the intervening valleys, these features are very striking objects when viewed under good conditions with high powers. In some cases they may possibly represent the effects of the slipping of the upper portions of the wall, from a want of cohesiveness in the material of which it is composed; but this hardly explains why the highest terrace often stands nearly as high as the rampart. Nasmyth, in his eruption hypothesis, suggests that in such a case there may have been two eruptions from the same vent; one powerful, which formed the exterior circle, and a second, rather less powerful, which has formed the interior circle. Ultimately, however, coming to the conclusion that terraces, as a rule, are not due to any such freaks of the eruption, he ascribes them to landslips. In any case, we can hardly imagine that material standing at such a high angle of inclination as that forming the summit ridge of many of the ring-plains would not frequently slide down in great masses, and thus form irregular plateaus on the lower and flatter portions of the slope; but this fails to explain the symmetrical arrangement of the concentric terraces and intermediate valleys. The inner declivity of the north-eastern wall of Plato exhibits what to all appearance is an undoubted landslip, as does also that of Hercules on the northern side, and numerous other cases might be adduced; but in all of them the appearance is very different from that of the true terrace.

The glacis, or outer slope of a ring-plain, is invariably of a much gentler inclination than that which characterises the inner declivity: while the latter very frequently descends at an angle varying from 60 deg. to 50 deg. at the crest of the wall, to from 10 deg. to 2 deg. at the bottom, where it meets the floor; the former extends for a great distance at a very flat gradient before it sinks to the general level of the surrounding country. It differs likewise from the inner descent, in the fact that, though often traversed by valleys, intersected by deep gullies and irregular depressions, and covered with humpy excrescences and craters, it is only rarely that any features comparable to the terraces, usually present on the inner escarpment, can be traced upon it.

Elongated depressions of irregular outline, and very variable in size and depth, are frequently found on the outer slopes of the border. Some of them consist of great elliptical or sub-circular cavities, displaying many expansions and contractions, called "pockets," and suggesting the idea that they were originally distinct cup-shaped hollows, which from some cause or other have coalesced like rows of inosculating craters. While many of these features are so deep that they remain visible for a considerable time under a low sun, others, though perhaps of greater extent, vanish in an hour or so.

As in the case of the walled-plains, the ramparts of the ring-plains exhibit gaps and are broken by craters and depressions, but to a much less extent. Often the lofty crest, surmounted by aiguilles or by blunter peaks, towering in some cases to nearly double its altitude above the interior, is perfectly continuous (like Copernicus), or only interrupted by narrow passes. It is a suggestive circumstance that gaps, other than valleys, are almost invariably found either in the north or south walls, or in both, and seldom in other positions. The buttress, or long-extending spur, is a feature frequently associated with the ring- plain rampart, as are also numbers of what, for the lack of a better name, must be termed little hillocks, which generally radiate in long rows from the outer foot of the slope. The spurs usually abut on the wall, and, either spreading out like the sticks of a fan or running roughly parallel to each other, extend for long distances, gradually diminishing in height and width till they die out on the surrounding surface. They have been compared to lava streams, which those round Aristillus, Aristoteles, and on the flank of Clavius a, certainly somewhat resemble, though, in the two former instances, they are rather comparable to immense ridges. In addition to the above, the spurs radiating from the south-eastern rampart of Condamine and the long undulating ridges and rows of hillocks running from Cyrillus over the eastern glacis of Theophilus, may be named as very interesting examples.

Neison and some other selenographers place in a distinct class certain of the smaller ring-plains which usually have a steeper outer slope, and are supposed to present clearer indications of a volcanic origin than the ring-plains, terming them "Crater-plains."

CRATERS.--Under this generic name is placed a vast number of formations exhibiting a great difference in size and outward characteristics, though generally (under moderate magnification) of a circular or sub-circular shape. Their diameter varies from 15 miles or more to 3, and even less, and their flanks rise much more steeply to the summit, which is seldom very lofty, than those of the ring-plains, and fall more gradually to the floor. There is no portion of the moon in which they do not abound, whether it be on the ramparts, floors, and outer slopes of walled and ring plains, the summits and escarpments of mountain ranges, amid the intricacies of the highlands, or on the grey surface of the Maria. In many instances they have a brighter and newer aspect than the larger formations, often being the most brilliant points on their walls, when they are found in this position. Very frequently too they are not only very bright themselves, but stand on bright areas, whose borders are generally concentric with them, which shine with a glistening lustre, and form a kind of halo of light around them. Euclides and Bessarion A, and the craters east of Landsberg, are especially interesting examples. It seems not improbable that these areas may represent deposits formed by some kind of matter ejected from the craters, but whether of ancient or modern date, it is, of course, impossible to determine. Future observers will perhaps be in a better position to decide the question without cavil, if such eruptions should again take place. Like the larger enclosures, these smaller objects frequently encroach upon each other-- crater-ring overlapping crater-ring, as in the case of Thebit, where a large crater, which has interfered with the continuity of the east wall, has, in its turn, been disturbed by a smaller crater on its own east wall. The craters in many cases, possibly in the majority if we could detect them, have central mountains, some of them being excellent tests for telescopic definition--as, for example, the central peaks of Hortensius, Bessarion, and that of the small crater just mentioned on the east wall of Thebit A. A tendency to a linear arrangement is often displayed, especially among the smaller class, as is also their occurrence in pairs.

CRATER-CONES.--These objects, plentifully distributed on the lunar surface, are especially interesting from their outward resemblance to the parasitic cones found on the flanks of terrestrial volcanoes (Etna, for instance). In the larger examples it is occasionally possible to see that the interiors are either inverted cones without a floor, or cup-shaped depressions on the summit of the object. Frequently, however, they are so small that the orifice can only be detected under oblique illumination. Under a high sun they generally appear as white spots, more or less ill-defined, as on the floors of Archimedes, Fracastorius, Plato, and many other formations, which include a great number, all of which are probably crater cones, although only a few have been seen as such. It is a significant fact that in these situations they are always found to be closely associated with the light streaks which traverse the interior of the formations, standing either on their surface or close to their edges. The instrumental and meteorological requirements necessary for a successful scrutiny of the smallest type of these features, are beyond the reach of the ordinary observer in this country, as they demand direct observation in large telescopes under the best atmospheric conditions.

Some years ago Dr. Klein of Cologne called attention to some very interesting types of crater-cones, which may be found on certain dark or smoky-grey areas on the moon. These, he considers, may probably represent active volcanic vents, and urges that they should be diligently examined and watched by observers who possess telescopes adequate to the task. The most noteworthy examples of these objects are in the following positions:--(1) West of a prominent ridge running from Beaumont to the west side of Theophilus, and about midway between these formations; (2) in the Mare Vaporum, south of Hyginus; (3) on the floor of Werner, near the foot of the north wall; (4) under the east wall of Alphonsus, on the dusky patch in the interior; (5) on the south side of the floor of Atlas. I have frequently described elsewhere with considerable detail the telescopic appearance of these features under various phases, and have pointed out that though large apertures and high powers are needed to see these cones to advantage, the dusky areas, easily traced on photograms, might be usefully studied by observers with smaller instruments, as if they represent the ejecta from the crater-cones which stand upon them, changes in their form and extent could very possibly be detected. In addition to those already referred to, a number of mysterious dark spots were discovered by Schmidt in the dusky region about midway between Copernicus and Gambart, which Klein describes as perforated like a sieve with minute craters. A short distance south-west of Copernicus stands a bright crater-cone surrounded by a grey nimbus, which may be classed with these objects. It is well seen under a high light, as indeed is the case with most of these features.

CRATERLETS, CRATER-PITS.--To a great extent the former term is needless and misleading, as the so-called craters merge by imperceptible gradations into very minute objects, as small as half a mile in diameter, and most probably, if we could more accurately estimate their size, still less. The crater-pit, however, has well-marked peculiarities which distinguish it from all other types, such as the absence of a distinguishable rim and extreme shallowness. They appear to be most numerous on the high-level plains and plateaus in the south-western quadrant, and may be counted by hundreds under good atmospheric conditions on the outer slopes of Walter, Clavius, and other large enclosures. In these positions they are often so closely aggregated that, as Nasmyth remarks, they remind one of an accumulation of froth. Even in an 8 1/2 inch reflector I have frequently seen the outer slope of the large ring-plain on the north-western side of Vendelinus, so perforated with these objects that it resembled pumice or vesicular lava, many of the little holes being evidently not circular, but square shaped and very irregular. The interior of Stadius and the region outside abounds in these minute features, but the well-known crater-row between this formation and Copernicus seems rather to consist of a number of inosculating crater-cones, as they stand very evidently on a raised bank of some altitude.

MOUNTAIN RANGES, ISOLATED MOUNTAINS, &c.--The more massive and extended mountain ranges of the moon are found in the northern hemisphere, and (what is significant) in that portion of it which exhibits few indications of other superficial disturbances. The most prominently developed systems, the Alps, the Caucasus, and the Apennines, forming a mighty western rampart to the Mare Imbrium and giving it all the appearance of a vast walled plain, present few points of resemblance to any terrestrial chain. The former include many hundred peaks, among which, Mont Blanc rises to a height of 12,000 feet, and a second, some distance west of Plato, to nearly as great an altitude; while others, ranging from 5000 to 8000 feet, are common. They extend in a south-west direction from Plato to the Caucasus, terminating somewhat abruptly, a little west of the central meridian, in about N. lat. 42 deg. One of the most interesting features associated with this range is the so-called great Alpine valley, which cuts through it west of Plato. The Caucasus consist of a massive wedge-shaped mountain land, projecting southwards, and partially dividing the Mare Imbrium from the Mare

Serenitatis, both of which they flank. Though without peaks so lofty as those pertaining to the Alps, there is one, immediately east of the ring-plain Calippus, which, towering to 19,000 feet, surpasses any of which the latter system can boast. The Apennines, however, are by far the most magnificent range on the visible surface, including as they do some 3000 peaks, and extending in an almost continuous curve of more than 400 miles in length from Mount Hadley, on the north, to the fine ring-plain Eratosthenes, which forms a fitting termination, on the south. The great headland Mount Hadley rises more than 15,000 feet, while a neighbouring promontory on the south-east of it is fully 14,000 feet, and another, close by, is still higher above the Mare. Mount Huygens, again, in N. lat. 20 deg., and the square-shaped mass Mount Wolf, near the southern end of the chain, include peaks standing 18,000 and 12,000 feet respectively above the plain, to which their flanks descend with a steep declivity. The counterscarp of the Apennines, in places 160 miles in width from east to west, runs down to the Mare Vaporum with a comparatively gentle inclination. It is everywhere traversed by winding valleys of a very intricate type, all trending towards the south-west, and includes some bright craters and mountain-rings. The Carpathians, forming in part the southern border of the Mare Imbrium, extend for a length of more than 180 miles eastward of E., long. 16 deg., and, embracing the ring-plain Gay- Lussac, terminate west of Mayer. They present a less definite front to the Mare than the Apennines, and are broken up and divided by irregular valleys and gaps; their loftiest peak, situated on a very projecting promontory north-west of Mayer, rising to a height of 7000 feet. Notwithstanding their comparatively low altitude, the region they occupy forms a fine telescopic picture at lunar sunrise. The _Sinus Iridum highlands_, bordering the beautiful bay on the north-east side of the Mare Imbrium, rank among the loftiest and most intricate systems on the moon, and, like the Apennines, present a steep face to the grey plain from which they rise, though differing from them in other respects. They include many high peaks, the loftiest, in the neighbourhood of the ring- plain Sharp, rising 15,000 feet. There are probably some still higher mountains in the vicinity, but the difficulties attending their measurement render it impossible to determine their altitude with any approach to accuracy.

The Taurus Mountains extend from the west side of the Mare Serenitatis, near Le Monnier and Littrow, in a north-westerly direction towards Geminus and Berselius, bordering the west side of the Lacus Somniorum. They are a far less remarkable system than any of the preceding, and consist rather of a wild irregular mountain region than a range. In the neighbourhood of Berselius are some peaks which, according to Neison, cannot be less than 10,000 feet in height.

On the north side of the Mare Imbrium, east of Plato, there is a beautiful narrow range of bright outlying heights, called the _Teneriffe Mountains_, which include many isolated objects of considerable altitude, one of the loftiest rising about 8000 feet. Farther towards the east lies another group of a very similar character, called the Straight Range, from its linear regularity. It extends from west to east for a distance of about 60 miles, being a few miles shorter than the last, and includes a peak of 6000 feet.

The Harbinger Mountains.--A remarkable group, north-west of Aristarchus, including some peaks as high as 7000 feet, and other details noticed in the catalogue.

The above comprise all the mountain ranges in the northern hemisphere of any prominence, or which have received distinctive names, except the Hercynian Mountains, on the north-east limb, east of the walled plain Otto Struve. These are too near the edge to be well observed, but, from what can be seen of them, they appear to abound in lofty peaks, and to bear more resemblance to a terrestrial chain than any which have yet been referred to.

The mountain systems of the southern hemisphere, except the ranges visible on the limb, are far less imposing and remarkable than those just described. The Pyrenees, on the western side of the Mare Nectaris, extend in a meridional direction for nearly 190 miles, and include a peak east of Guttemberg of nearly 12,000 feet, and are traversed in many places by fine valleys.

The Altai Mountains form a magnificent chain, 275 miles in length, commencing on the outer eastern slope of Piccolomini, and following a tolerably direct north-east course, with a few minor bendings, to the west side of Fermat, where they turn more towards the north, ultimately terminating about midway between Tacitus and Catherina. The region situated on the south-east is a great table-land, without any prominent features, rising gently towards the mountains, which shelve steeply down to an equally barren expanse on the north-west, to which they present a lofty face, having an average altitude of about 6000 feet. The loftiest peak, over 13,000 feet, rises west of Fermat.

The Riphaean Mountains, a remarkably bright group, occupying an isolated position in the Mare Procellarum south of Landsberg, and extending for more than 100 miles in a meridional direction. They are most closely aggregated at a point nearly due west of Euclides, from which they throw off long-branching arms to the north and south, those on the north bifurcating and gradually sinking to the level of the plain. The loftiest peaks are near the extremity of this section, one of them rising to 3000 feet. Two bright craters are associated with these mountains, one nearly central, and the other south of it.

The Percy Mountains.--This name is given to the bright highlands extending east of Gassendi towards Mersenius, forming the north-eastern border of the Mare Humorum. They abound in minute detail--bright little mountains and ridges--and include some clefts pertaining to the Mersenius rill-system; but their most noteworthy feature is the long bright mountain-arm, branching out from the eastern wall of Gassendi, and extending for more than 50 miles towards the south-east.

The principal ranges on the limb are the Leibnitz Mountains, extending from S. lat. 70 deg. on the west to S. lat. 80 deg. on the east limb. They include some giant peaks and plateaus, noteworthy objects in profile, some of which, according to Schroter and Madler, rise to 26,000 feet. The Dorfel Mountains, between S. lat. 80 deg. and 57 deg. on the eastern limb, include, if Schroter's estimate is correct, three peaks which exceed 26,000 feet. On the eastern limb, between S. lat. 35 deg. and 18 deg., extend the Rook Mountains, which have peaks, according to Schroter, as high as 25,000 feet. Next in order come the Cordilleras, which extend to S. lat. 8 deg., and the D'Alembert Mountains, lying east of Rocca and Grimaldi, closely associated with them, and probably part of the same system. Some of the peaks approach 20,000 feet. In addition to these mountain ranges there are others less prominent on the limb in the northern hemisphere, which have not been named.

ISOLATED MOUNTAINS are very numerous in different parts of the moon, the most remarkable are referred to in the appendix. Many remain unnamed.

CLEFTS OR RILLS.--Though Fontenelle, in his _Entretiens sur la Pluralite des Mondes_, informs his pupil, the Marchioness, that "M. Cassini discovered in the moon something which separates, then reunites, and finally loses itself in a cavity, which from its appearance seemed to be a river," it can hardly be supposed that what the French astronomer saw, or fancied he saw, with the imperfect telescopes of that day, was one of the remarkable and enigmatical furrows termed clefts or rills, first detected by the Hanoverian selenographer Schroter; who, on October 7, 1787, discovered the very curious serpentine cleft near Herodotus, having a few nights before noted for the first time the great Alpine valley west of Plato, once classed with the clefts, though it is an object of a very different kind. Between 1787 and 1797 Schroter found ten rills; but twenty years elapsed before an addition was made to this number by the discoveries of Gruithuisen, and, a short time after, by those of Lohrmann, who in twelve months (1823-24) detected seventy. Kinau, Madler, and finally Schmidt, followed, till, in 1866, when the latter published his work, Ueber Rillen auf dem Monde, the list was thus summarised:--

In the 1st or N.W. quadrant 127 rills In the 2nd or N.E. quadrant 75 rills In the 3rd or S.E. quadrant 141 rills In the 4th or S.W. quadrant 82 rills

or 425 in all. Since the date of this book the number of known rills has been more than doubled; in fact, scarcely a lunation passes without new discoveries being made.

The significance of the word rille in German, a groove or furrow, describes with considerable accuracy the usual appearance of the objects to which it is applied, consisting as they do of long narrow channels, with sides more or less steep, and sometimes vertical. They often extend for hundreds of miles in approximately straight lines over portions of the moon's surface, frequently traversing in their course ridges, craters, and even more formidable obstacles, without any apparent check or interruption, though their ends are sometimes marked by a mound or crater. Their length ranges from ten or twelve to three hundred miles or more (as in the great Sirsalis rill), their breadth, which is very variable within certain limits, from less than half a mile to more than two, and their depth (which must necessarily remain to a great extent problematical) from 100 to 400 yards. They exhibit in the telescope a gradation from somewhat coarse grooves, easily visible at suitable times in very moderately sized instruments, to striae so delicate as to require the largest and most perfect optical means and the best atmospheric conditions to be glimpsed at all. Viewed under moderate amplification, the majority of rills resemble deep canal-like channels with roughly parallel sides, displaying occasionally local irregularities, and fining off to invisibility at one or both ends. But, if critically scrutinised in the best observing weather with high powers, the apparent evenness of their edges entirely disappears, and we find that the latter exhibit indentations, projections, and little flexures, like the banks of an ordinary stream or rivulet, or, to use a very homely simile, the serrated edges and little jagged irregularities of a biscuit broken across. In some cases we remark crateriform hollows or sudden expansions in their course, and deep sinuous ravines, which render them still more unsymmetrical and variable in breadth. With regard to their distribution on the lunar surface; they are found in almost every region, but perhaps not so frequently on the surface of the Maria as elsewhere, though, as in the case of the Triesnecker and other systems, they often abound in the neighbourhood of disturbed regions in these plains, and in many cases along their margins, as, for example, the Gassendi-Mersenius and the Sabine-Ritter groups. The interior of walled plains are frequently intersected by them, as in Gassendi, where nearly forty, more or less delicate examples, have been seen; in Hevel, where there is a very interesting system of crossed clefts, and within Posidonius. If we study any good modern lunar map, it is evident how constantly they appear near the borders of mountain ranges, walled-plains, and ring-plains; as, for instance, at the foot of the Apennines; near Archimedes, Aristarchus, Ramsden, and in many other similar positions. Rugged highlands also are often traversed by them, as in the case of those lying west of Le Monnier and Chacornac, and in the region west of the Mare Humorum. It may be here remarked, however, as a notable fact, that the neighbourhood of the grandest ring-mountain on the moon, Copernicus, is, strange to say, devoid of any features which can be classed as true clefts, though it abounds in crater-rows. The intricate network of rills on the west of Triesnecker, when observed with a low power, reminds one of the wrinkles on the rind of an orange or on the skin of a withered apple. Gruithuisen, describing the rill-traversed region between Agrippa and Hyginus, says that "it has quite the look of a Dutch canal map." In the subjoined catalogue many detailed examples will be given relating to the course of these mysterious furrows; how they occasionally traverse mountain arms, cut through, either completely or partially (as in Ramsden), the borders of ring-plains and other enclosures, while not unfrequently a small mound or similar feature appears to have caused them to swerve suddenly from their path, as in the case of the Ariadaeus cleft, and in that of one member of the Mercator-Campanus system.

Of the actual nature of the lunar rills we are, it must be confessed, supremely ignorant. With some of the early observers it was a very favourite notion that they are artificial works, constructed presumably by Kepler's sub-volvani, or by other intelligences. There is perhaps some excuse to be made for the freaks of an exuberant fancy in regard to objects which, if we ignore for a moment their enormous dimensions, judged by a terrestrial standard, certainly have, in their apparent absence of any physical relation to neighbouring objects, all the appearance of being works of art rather than of nature. The keen-sighted and very imaginative Gruithuisen believed that in some instances they represent roads cut through interminable forests, and in others the dried-up beds of once mighty rivers. His description of the Triesnecker rill-system reads like a page from a geographical primer. A portion of it is compared to the river Po, and he traces its course mile by mile up to the "delta" at its place of disemboguement into the Mare Vaporum. From the position of some rills with respect to the contour of the surrounding country, it is evident that if water were now present on the moon, they, being situated at the lowest level, would form natural channels for its reception; but the exceptions to this arrangement are so numerous and obvious, that the idea may be at once dismissed that there is any analogy between them and our rivers. The eminent selenographer, the late W.K. Birt, compared many of them to "inverted river-beds" from the fact that, as often as not, they become broader and deeper as they attain a higher level. The branches resemble rivers more frequently than the main channels; for they generally commence as very fine grooves, and, becoming broader and broader, join them at an acute angle. An attempt again has been made to compare the lunar clefts with those vast gorges, the marvellous results of aqueous action, called canyons, which attain their greatest dimensions in North America; such as the Great Canyon of the Colorado, which is at least 300 miles in length, and in places 2000 yards in depth, with perpendicular or even overhanging sides; but the analogy, at first sight specious, utterly breaks down under closer examination. Some selenographers consider them to consist of long-extending rows of confluent craters, too minute to be separately distinguished, and to be thus due to some kind of volcanic action. This is undoubtedly true in many instances, for almost every lunar region affords examples of crater-rows merging by almost imperceptible gradations into cleft-like features, and crater-rows of considerable size resemble clefts under low powers. Still it seems probable that the greater number of these features are immense furrows or cracks in the surface and nothing more; for the higher the magnifying power employed in their examination, the less reason there is to object to this description. Dr. Klein of Cologne believes that rills of this class are due to the shrinkage of parts of the moon's crust, and that they are not as a rule the result of volcanic causes, though he admits that there may be some which have a seismic origin. No good reason has as yet been given for the fact that they so frequently cross small craters and other objects in their course, though it has been suggested that the route followed by a rill from crater to crater in these instances may be a line of least surface resistance, an explanation far from being satisfactory.

Whether variations in the visibility of lunar details, when observed under apparently similar conditions, actually occur from time to time from some unknown cause, is one of those vexed questions which will only be determined when the moon is systematically studied by experienced observers using the finest instruments at exceptionally good stations; but no one who examines existing records of observations of rills by Gruithuisen, Lohrmann, Madler, Schmidt, and other observers, can well avoid the conclusion that the anomalies brought to light therein point strongly to the probability of the existence of some agency which occasionally modifies their appearance or entirely conceals them from view.

The following is one illustration out of many which might be quoted. At a point in its course, nearly due north of the ring-plain Agrippa, the great Ariadaeus cleft sends out a branch which runs into the well-known Hyginus cleft, reminding one, as Dr. Klein remarks, of two rivers connected in the shortest way by a canal. This uniting furrow was detected by Gruithuisen, who observed it several times. On some occasions it appeared perfectly straight, at others very irregular; but, what is very remarkable, although two such accurate observers as Lohrmann and Madler frequently scrutinised the region, neither of them saw a trace of this object; and but for its rediscovery by Schmidt in 1862, its existence would certainly have been ignored by selenographers as a mere figment of Gruithuisen's too lively imagination. Dr. Klein has frequently seen this rill with great distinctness, and at other times sought for it in vain; though on each occasion the conditions of illumination, libration, and definition were practically similar. I have sometimes found this cleft an easy object with a 4 inch achromatic. Again, many rills described by Madler as very delicate and difficult to trace, may now be easily followed in "common telescopes." In short, the more direct telescopic observations accumulate, and the more the study of minute detail is extended, the stronger becomes the conviction, that in spite of the absence of an appreciable atmosphere, there may be something resembling low-lying exhalations from some parts of the surface which from time to time are sufficiently dense to obscure, or even obliterate, the region beneath them.

If, as seems most probable, these gigantic cracks are due to contractions of the moon's surface, it is not impossible, in spite of the assertions of the textbooks to the effect that our satellite is now "a changeless world," that emanations may proceed from these fissures, even if, under the monthly alternations of extreme temperatures, surface changes do not now occasionally take place from this cause also. Should this be so, the appearance of new rills and the extension and modification of those already existing may reasonably be looked for. Many instances might be adduced tending to confirm this supposition, to one of which, as coming under my notice, I will briefly refer. On the evening of November 11, 1883, when examining the interior of the great ring-plain Mersenius with a power of 350 on an 8 1/2 inch reflector; in addition to the two closely parallel clefts discovered by Schmidt, running from the inner foot of the north-eastern rampart towards the centre, I remarked another distinct cleft crossing the northern part of the floor from side to side. Shortly afterwards, M. Gaudibert, one of our most experienced selenographers, who has discovered many hitherto unrecorded clefts, having seen my drawing, searched for this object, and, though the night was far from favourable, had distinct though brief glimpses of it with the moderate magnifying power of 100. Mersenius is a formation about 40 miles in diameter, with a prominently convex interior, containing much detail which has received more than ordinary attention from observers. It has, moreover, been specially mapped by Schmidt and others, yet no trace of this rill was noted, though objects much more minute and difficult have not been overlooked. Does not an instance of this kind raise a well-grounded suspicion of recent change which it is difficult to explain away?

To see the lunar clefts to the best advantage, they must be looked for when not very far removed from the terminator, as when so situated the black shadow of one side, contrasted with the usually brightly- illuminated opposite flank, renders them more conspicuous than when they are viewed under a higher sun. Though, as a rule, invisible at full moon, some of the coarser clefts--as, for example, a portion of the Hyginus furrow, and that north of Birt--may be traced as delicate white lines under a nearly vertical light.

For properly observing these objects, a power of not less than 300 on telescopes of large aperture is needed; and in studying their minute and delicate details, we are perhaps more dependent on atmospheric conditions than in following up any other branch of observational astronomy. Few indeed are the nights, in our climate at any rate, when the rough, irregular character of the steep interior of even the coarser examples of these immense chasms can be steadily seen. We can only hope to obtain a more perfect insight into their actual structural peculiarities when they are scrutinised under more perfect climatic circumstances than they have been hitherto. When observing the Hyginus cleft, Dr. Klein noticed that at one place the declivities of the interior displayed decided differences of tint. At many points the reflected sunlight was of a distinctly yellow hue, while in other places it was white, as if the cliffs were covered with snow. He compares this portion of the rill to the Rhine valley between Bingen and Coblentz, but adds that the latter, if viewed from the moon, would probably not present so fresh an appearance, and would, of course, be frequently obscured by clouds.

Since the erection of the great Lick telescope on Mount Hamilton, our knowledge of the details of some of the lunar clefts has been greatly increased, as in the case of the Ariadaeus cleft, and many others. Professor W.H. Pickering, also, at Arequipa, has made at that ideal astronomical site many observations which, when published, will throw more light upon their peculiar characteristics.

A few years ago M.E.L. Trouvelot of Meudon drew attention to a curious appearance which he noted in connection with certain rills when near the terminator, viz., extremely attenuated threads of light on their sites and their apparent prolongations. He observed it in the ring-plain Eudoxus, crossing the southern side of the floor from wall to wall; and also in connection with the prominent cleft running from the north side of Burg to the west of Alexander, and in some other situations. He terms these phenomena Murs enigmatiques.

Apparent prolongations of clefts in the form of rows of hillocks or small mounds are very common.

FAULTS.--These sudden drops in the surface, representing local dislocations, are far from unusual: the best examples being the straight wall, or "railroad," west of Birt; that which strikes obliquely across Plato; another which traverses Phocylides; and a fourth that has manifestly modified the mountain arm north of Cichus. They differ from the terrestrial phenomena so designated in the fact that the surface indications of these are destroyed by denudation or masked by deposits of subsequent date. In many cases on the moon, though its course cannot be traced in its entirety by its shadow, yet the existence of a fault may be inferred by the displacement and fracture of neighbouring objects.

VALLEYS.--Features thus designated, differing greatly both in size and character, are met with in almost every part of the surface, except on the grey plains. While the smallest examples, from their delicacy, tenuity, and superficial resemblance to rills, are termed rill-valleys, the larger and more conspicuous assume the appearance of coarse chasms, gorges, or trough-like depressions. Between these two extremes, are many objects of moderate dimensions--winding or straight ravines and defiles bounded by steep mountains, and shallow dales flanked by low rounded heights. The rill valleys are very numerous, only differing in many instances from the true rills in size, and are probably due to the same cause. Among the most noteworthy valleys of the largest class must, of course, be placed the great valley of the Alps, one of the most striking objects in the northern hemisphere, which also includes the great valley south-east of Ukert. The Rheita valley, the very similar chasm west of Reichenbach, and the gorge west of Herschel, are also notable examples in the southern hemisphere. The borders of some of the Maria (especially that of the Mare Crisium) and of many of the depressed rimless formations, furnish instances of winding valleys intersecting their borders: the hilly regions likewise often abound in long branching defiles.

BRIGHT RAY-SYSTEMS.--Reference has already been made to the faint light streaks and markings often found on the floors of the ring-mountains and in other situations, and to the brilliant nimbi surrounding some of the smaller craters; but, in addition to these, many objects on the moon's visible surface are associated with a much more remarkable and conspicuous phenomenon--

the bright rays which, under a high sun, are seen either to radiate from them as apparent centres to great distances, or, in the form of irregular light areas, to environ them, and to throw out wide-spreading lucid beams, extending occasionally many hundreds of miles from their origin. The more striking of these systems were recognised and drawn at a very early stage of telescopic observation, as may be seen if we consult the quaint old charts of Hevel, Riccioli, Fontana, and other observers of the seventeenth century, where they are always prominently, though very inaccurately, portrayed. The principal ray-systems are those of Tycho, Copernicus, Kepler, Anaxagoras, Aristarchus, Olbers, Byrgius A, and Zuchius; while Autolycus, Aristillus, Proclus, Timocharis, Furnerius A, and Menelaus are grouped as constituting minor systems. Many additional centres exist, a list of which will be found in the appendix.

The rays emanating from Tycho surpass in extent and interest any of the others. Hundreds of distinct light streaks originate round the grey margin of this magnificent object, some of them extending over a greater part of the moon's visible superficies, and "radiating," in the words of Professor Phillips, "like false meridians, or like meridians true to an earlier pole of rotation." No systematic attempt has yet been made to map them accurately as a whole on a large scale, for their extreme intricacy and delicacy would render the task a very difficult one, and, moreover, would demand a long course of observation with a powerful telescope in an ideal situation; but Professor W.H. Pickering, observing under these conditions at Arequipa, has recently devoted considerable attention both to the Tycho and other rays, with especially suggestive and important results, which may be briefly summarised as follows:-

(1.) That the Tycho streaks do not radiate from the apparent centre of this formation, but point towards a multitude of minute craterlets on its south-eastern or northern rims. Similar craterlets occur on the rims of other great craters, forming ray-centres. (2.) Speaking generally, a very minute and brilliant crater is located at the end of the streak nearest the radiant point, the streak spreading out and becoming fainter towards the other end. The majority of the streaks appear to issue from one or more of these minute craters, which rarely exceed a mile in diameter. (3.) The streaks which do not issue from minute craters, usually lie upon or across ridges, or in other similar exposed situations, sometimes apparently coming through notches in the mountain walls. (4.) Many of the Copernicus streaks start from craterlets within the rim, flow up the inside and down the outside of the walls. Kepler includes two such craterlets, but here the flow seems to have been more uniform over the edges of the whole crater, and is not distinctly divided up into separate streams. (5.) Though there are similar craters within Tycho, the streaks from them do not extend far beyond the walls. All the conspicuous Tycho streaks originate outside the rim. (6.) The streaks of Copernicus, Kepler, and Aristarchus are greyish in colour, and much less white than those associated with Tycho: some white lines extending south-east from Aristarchus do not apparently belong to the system. In the case of craterlets lying between Aristarchus and Copernicus the streaks point away from the latter. (7.) There are no very long streaks; they vary from ten to fifty miles in length, and are rarely more than a quarter of a mile broad at the crater. From this point they gradually widen out and become fainter. Their width, however, at the end farthest from the crater, seldom exceeds five miles.

These statements, especially those relating to the length of the streaks, are utterly opposed to prevailing notions, but Professor Pickering specifies the case of the two familiar parallel rays extending from the north-east of Tycho to the region east of Bullialdus. His observations show that these streaks, originating at a number of little craters situated from thirty to sixty miles beyond the confines of Tycho, "enter a couple of broad slightly depressed valleys. In these valleys are found numerous minute craters of the kind above described, with intensely brilliant interiors. When the streaks issuing from those craters near Tycho are nearly exhausted, they are reinforced by streaks from other craters which they encounter upon the way, the streaks becoming more pronounced at these points. These streaks are again reinforced farther out. These parallel rays must therefore not be considered as two streaks, but as two series of streaks, the components of which are placed end to end."

Thus, according to Professor Pickering, we must no longer regard the rays emanating from the Tycho region and other centres as continuous, but as consisting of a succession of short lengths, diminishing in brilliancy but increasing in width, till they reach the next crater lying in their direction, when they are reinforced; and the same process of gradual diminution in brightness and reinforcement goes on from one end to the other.

The following explanation is suggested to account for the origin of the rays:-

-"The earth and her satellite may differ not so much as regards volcanic action as in the densities of their atmospheres. Thus if the craterlets on the rim of Tycho were constantly giving out large quantities of gas or steam, which in other regions was being constantly absorbed or condensed, we should have a wind uniformly blowing away from that summit in all directions. Should other summits in its vicinity occasionally give out gases, mixed with any fine white powder, such as pumice, this powder would be carried away from Tycho, forming streaks."

The difficulty surrounding this very ingenious hypothesis is, that though, assuming the existence of pumice-emitting craters and regions of condensation, there might be a more or less lineal and streaky deposition of this white material over large areas of the moon, why should this deposit be so definitely arranged, and why should these active little craters happen to lie on these particular lines?

The confused network of streaks round Copernicus seem to respond more happily to the requirements of Professor Pickering's hypothesis, for here there is an absence of that definiteness of direction so manifestly displayed in the case of the Tycho rays, and we can well imagine that with an area of condensation surrounding this magnificent object beyond the limits of the streaks, and a number of active little craters on and about its rim, the white material ejected might be drawn outwards in every direction by wind currents, which possibly once existed, and, settling down, assume forms such as we see.

Nasmyth's well-known hypothesis attributes the radiating streaks to cracks in the lunar globe caused by the action of an upheaving force, and accounts for their whiteness by the outwelling of lava from them which has spread to a greater or less distance on either side. If the moon has been fractured in this way, we can easily suppose that the craters formed on these fissures, being in communication with the interior, might eject some pulverulent white matter long after the rest of the surface with its other types of craters had attained a quiescent stage.

The Tycho rays, when viewed under ordinary conditions, appear to extend in unbroken bands to immense distances. One of the most remarkable, strikes along the eastern side of Fracastorius, across the Mare Nectaris to

Guttemberg, while another, more central, extends, with local variations in brightness, through Menelaus, over the Mare Serenitatis nearly to the north-west limb. This is the ray that figures so prominently in rude woodcuts of the moon, in which the Mare Serenitatis traversed by it is made to resemble the Greek letter PHI. The Kepler, Aristarchus, and Copernicus systems, though of much smaller extent, are very noteworthy from the crossing and apparent interference of the rays; while those near Byrgius, round Aristarchus, and the rays from Proclus, are equally remarkable.

[Nichol found that the rays from Kepler cut through rays from Copernicus and Aristarchus, while rays from the latter cut through rays from the former. He therefore inferred that their relative ages stand in the order,--Copernicus, Aristarchus, Kepler.]

As no branch of selenography has been more neglected than the observation of these interesting but enigmatical features, one may hope that, in spite of the exacting conditions as to situation and instrumental requirements necessary for their successful scrutiny, the fairly equipped amateur in this less favoured country will not be deterred from attempting to clear up some of the doubts and difficulties which at present exist as to their actual nature.

THE MOON'S ALBEDO, SURFACE BRIGHTNESS, &c.--Sir John Herschel maintained that "the actual illumination of the lunar surface is not much superior to that of weathered sandstone rock in full sunshine." "I have," he says, "frequently compared the moon setting behind the grey perpendicular facade of the Table Mountain, illuminated by the sun just risen in the opposite quarter of the horizon, when it has been scarcely distinguishable in brightness from the rock in contact with it. The sun and moon being at nearly equal altitudes, and the atmosphere perfectly free from cloud or vapour, its effect is alike on both luminaries." Zollner's elaborate researches on this question are closely in accord with the above observational result. Though he considers that the brightest parts of the surface are as white as the whitest objects with which we are acquainted, yet, taking the reflected light as a whole, he finds that the moon is more nearly black than white. The most brilliant object on the surface is the central peak of the ring-plain Aristarchus, the darkest the floor of Grimaldi, or perhaps a portion of that of the neighbouring Riccioli. Between these extremes, there is every gradation of tone. Proctor, discussing this question on the basis of Zollner's experiments respecting the light reflected by various substances, concludes that the dark area just mentioned must be notably darker than the dark grey syenite which figures in his tables, while the floor of Aristarchus is as white as newly fallen snow.

The estimation of lunar tints in the usual way, by eye observations at the telescope, involving as it does physiological errors which cannot be eliminated, is a method far too crude and ambiguous to form the basis of a scientific scale or for the detection of slight variations. An instrument on the principle of Dawes' solar eyepiece has been suggested; this, if used with an invariable and absolute scale of tints, would remove many difficulties attending these investigations. The scale which was adopted by Schroter, and which has been used by selenographers up to the present time, is as follows:--

0 deg. = Black. 1 deg. = Greyish black. 2 deg. = Dark grey. 3 deg. = Medium grey. 4 deg. = Yellowish grey. 5 deg. = Pure light grey. 6 deg. = Light whitish grey. 7 deg. = Greyish white. 8 deg. = Pure white. 9 deg. = Glittering white. 10 deg. = Dazzling white.

The following is a list of lunar objects published in the Selenographical Journal, classed in accordance with this scale:--

0 deg. Black shadows. 1 deg. Darkest portions of the floors of Grimaldi and Riccioli. 1 1/2 deg. Interiors of Boscovich, Billy, and Zupus. 2 deg. Floors of Endymion, Le Monnier, Julius Caesar, Cruger, and Fourier a. 2 1/2 deg. Interiors of Azout, Vitruvius, Pitatus, Hippalus, and Marius. 3 deg. Interiors of Taruntius, Plinius, Theophilus, Parrot, Flamsteed, and Mercator. 3 1/2 deg. Interiors of Hansen, Archimedes, and Mersenius. 4 deg. Interiors of Manilius, Ptolemaeus, and Guerike. 4 1/2 deg. Surface round Aristillus, Sinus Medii. 5 deg. Walls of Arago, Landsberg, and Bullialdus. Surface round Kepler and Archimedes. 5 1/2 deg. Walls of Picard and Timocharis. Rays from Copernicus. 6 deg. Walls of Macrobius, Kant, Bessel, Mosting, and Flamsteed. 6 1/2 deg. Walls of Langrenus, Theaetetus, and Lahire. 7 deg. Theon, Ariadaeus, Bode B, Wichmann, and Kepler. 7 1/2 deg. Ukert, Hortensius, Euclides. 8 deg. Walls of Godin, Bode, and Copernicus. 8 1/2 deg. Walls of Proclus, Bode A, and Hipparchus c. 9 deg. Censorinus, Dionysius, Mosting A, and Mersenius B and c.

9 1/2 deg. Interior of Aristarchus, La Peyrouse DELTA. 10 deg. Central peak of Aristarchus.

TEMPERATURE OF THE MOON'S SURFACE.--Till the subject was undertaken some years ago by Lord Rosse, no approach was made to a satisfactory determination of the surface temperature of the moon. From his experiments he inferred that the maximum temperature attained, at or near the equator, about three days after full moon, does not exceed 200 deg. C., while the minimum is not much under zero C. Subsequent experiments, however, both by himself and Professor Langley, render these results more than doubtful, without it is admitted that the moon has an atmospheric covering. Langley's results make it probable that the temperature never rises above the freezing-point of water, and that at the end of the prolonged lunar night of fourteen days it must sink to at least 200 deg. below zero. Mr. F.W. Verey of the Alleghany Observatory has recently conducted, by means of the bolometer, similar researches as to the distribution of the moon's heat and its variation with the phase, by which he has deduced the varying radiation from the surface in different localities of the moon under various solar altitudes.

LUNAR OBSERVATION.--In observing the moon, we enjoy an advantage of which we cannot boast when most other planetary bodies are scrutinised; for we see the actual surface of another world undimmed by palpable clouds or exhalations, except such as exist in the air above us; and can gaze on the marvellous variety of objects it presents much as we contemplate a relief map of our own globe. But inasmuch as the manifold details of the relief map require to be placed in a certain light to be seen to the best advantage, so the ring-mountains, rugged highlands, and wide-extending plains of our satellite, as they pass in review under the sun, must be observed when suitable conditions of illumination prevail, if we wish to appreciate their true character and significance.

As a general rule, lunar objects are best seen when they are at no great distance from "the terminator," or the line dividing the illumined from the unillumined portion of the spherical surface. This line is constantly changing its position with the sun, advancing slowly onwards towards the east at a rate which, roughly speaking, amounts to about 30.5 min. in an hour, or passing over 10 deg. of lunar longitude in about 19 hrs. 40 mins. When an object is situated on this line, the sun is either rising or setting on the neighbouring region, and every inequality of the surface is rendered prominent by its shadow; so that trifling variations in level and minor asperities assume for the time being an importance to which they have no claim. If we are observing an object at lunar sunrise, a very short time, often only a few minutes, elapses before the confusion caused by the presence of the shadows of these generally unimportant features ceases to interfere with the observation, and we can distinguish between those details which are really noteworthy and others which are trivial and evanescent. Every formation we are studying should be observed, and drawn if possible, under many different conditions of illumination. It ought, in fact, to be examined from the time when its loftiest heights are first illumined by the rising sun till they disappear at sunset. This is, of course, practically impossible in the course of one lunation, but by utilising available opportunities, a number of observations may be obtained under various phases which will be more or less exhaustive. It cannot be said that much is known about any object until an attempt has been made to carry out this plan. Features which assume a certain appearance at one phase frequently turn out to be altogether different when viewed under another; important details obscured by shadows, craters masked by those of neighbouring objects, or by the shadows of their own rims, are often only revealed when the sun has attained an altitude of ten degrees or more. In short, there is scarcely a formation on the moon which does not exemplify the necessity of noting its aspect from sunrise to sunset. Regard must also be had to libration, which affects to a greater or less degree every object; carrying out of the range of observation regions near the limb at one time, and at another bringing into view others beyond the limits of the maps, which represent the moon in the mean state of libration. The area, in fact, thus brought into view, or taken out of it, is between 1/12th and 1/13th of the entire area of the moon, or about the 1/6th part of the hemisphere turned away from the earth. It is convenient to bear in mind that we see an object under nearly the same conditions every 59 d. 1 h. 28 m., or still more accurately, after the lapse of fifteen lunations, or 442 d. 23 h. Many observers avoid the observation of objects under a high light. This, however, should never be neglected when practicable, though in some cases it is not easy to carry out, owing to the difficulty in tracing details under these circumstances.

Although to observe successfully the minuter features, such as the rills and the smaller craterlets, requires instruments of large aperture located in favourable situations, yet work of permanent value may be accomplished with comparatively humble telescopic means. A 4 inch achromatic, or a silver-on-glass reflector of 6 or 6 1/2 inches aperture, will reveal on a good night many details which have not yet been recorded, and the possessor of instruments of this size will not be long in discovering that the moon, despite of what is often said, has not been so exhaustively surveyed that nothing remains for him to do.

Only experience and actual trial will teach the observer to choose the particular eyepiece suitable for a given night or a given object. It will be found that it is only on very rare occasions that he can accomplish much with powers which, perhaps only on two or three nights in a year in this climate, tell to great advantage; though it sometimes happens that the employment of an eyepiece, otherwise unsuitable for the night, will, during a short spell of good definition, afford a fleeting glimpse of some difficult feature, and thus solve a doubtful point. It has often been said that the efficiency of a telescope depends to a great extent on "the man at the eye end." This is as true in the case of the moon as it is in other branches of observational astronomy.

Observers, especially beginners, frequently fall into great error in failing to appreciate the true character of what they see. In this way a shallow surface depression, possibly only a few feet below the general level of the neighbouring country, is often described as a "vast gorge," because, under very oblique light, it is filled with black shadow; or an insignificant hillock is magnified into a mountain when similarly viewed. Hence the importance, just insisted on, of studying lunar features under as many conditions as possible before finally attempting to describe them.

However indifferent a draughtsman an observer may be, if he endeavours to portray what he sees to the best of his ability, he will ultimately attain sufficient skill to make his work useful for future reference: in any case, it will be of more value than a mere verbal description without a sketch. Doubt and uncertainty invariably attend to a greater or less extent written notes unaccompanied by drawings, as some recent controversies, respecting changes in Linne and elsewhere, testify. Now that photographs are generally available to form the basis of a more complete sketch, much of the difficulty formerly attending the correct representation of the outline and grosser features of a formation has been removed, and the observer can devote his time and attention to the insertion and description of less obvious objects.

PROGRESS OF SELENOGRAPHY.--Till within recent years, the systematic study of the lunar surface may be said to have been confined, in this country at any rate, to a very limited number of observers, and, except in rare instances, those who possessed astronomical telescopes only directed them to the moon as a show object to excite the wonder of casual visitors. The publication of Webb's "Celestial Objects" in 1859, the supposed physical change in the crater Linne, announced in 1866, and the appearance of an unrecorded black spot near Hyginus some ten years later, had the effect of awakening a more lively interest in selenography, and undoubtedly combined to bring about a change in this respect, which ultimately resulted in the number of amateurs devoting much of their time to this branch of observational astronomy being notably increased. Still, large telescopes, as a rule, held aloof for some unexplained reason, or were only employed in a desultory and spasmodic fashion, without any very definite object. When the Council of the British Association for the Advancement of Science, stimulated by the Linne controversy, deemed the moon to be worthy of passing attention, observations, directed to objects suspected of change (the phenomena on the floor of Plato) were left to three or four observers, under the able direction of Mr. Birt, the largest instruments available being an 8 1/4 inch reflector and the Crossley refractor of 9 inches aperture! During the last decade, however, all this has been changed, and we not only have societies, such as the British Astronomical Association, setting apart a distinct section for the systematic investigation of lunar detail, but some of the largest and most perfect instruments in the world, among them the noble refractor on Mount Hamilton, employed in photographing the moon or in scrutinising her manifold features by direct observation. Hence, it may be said that selenography has taken a new and more promising departure, which, among other results, must lead to a more accurate knowledge of lunar topography, and settle possibly, ere long, the vexed question of change, without any residuum of doubt.

Lunar photography as exemplified by the marvellous and beautiful pictures produced at the Lick Observatory under the auspices of Dr. Holden, and the exquisite enlargements of them by Dr. Weinek of Prague; at Paris by the brothers Henry; and at Brussels by M. Prinz; point to the not far distant time when we shall possess complete photographic maps on a large scale of the whole visible disc under various phases of illumination, which will be of inestimable value as topographical charts. When this is accomplished, the observer will have at his command faithful representations of any formation, or of any given region he may require, to utilise for the study of the smaller details by direct observation.

Desultory and objectless drawings and notes have hitherto been more or less characteristic of the work done, even by those who have given more than ordinary attention to the moon. Though these, if duly recorded, are valuable as illustrating the physical structure, the estimated brightness under various phases, and other peculiarities of lunar features, they do not materially forward investigations relating to the discovery of present lunar activity or to the detection of actual change. It is reiterated ad nauseam in many popular books that the moon is a changeless world, and it is implied that, having attained a state when no further manifestations of internal or external forces are possible, it revolves round the earth in the condition, for the most part, of a globular mass of vesicular lava or slag, possessing no interest except as a notable example of a "burnt-out planet." In answer to these dogmatic assertions, it may be said that, notwithstanding the multiplication of monographs and photographs, the knowledge we possess, even of the larger and more prominent objects, is far too slight to justify us in maintaining that changes, which on earth we should use a strong adjective to describe, have not taken place in connection with some of them in recent years. Would the most assiduous observer assert that his knowledge of any one of the great formations, in the south-west quadrant, for example, is so complete that, if a chasm as big as the Val del Bove was blown out from its flanks, or formed by a landslip, he would detect the change in the appearance of an area (some three miles by four) thus brought about, unless he had previously made a very prolonged and exhaustive study of the object? Or, again, among formations of a different class, the craters and crater-cones; might not objects as large as Monte Nuovo or Jorullo come into existence in many regions without any one being the wiser? It would certainly have needed a persistent lunar astronomer, and one furnished with a very perfect telescope, to have noted the changes that have occurred within the old crater-ring of Somma or among the Santorin group during the past thirty years, or even to have detected the effects resulting from the great catastrophe in A.D. 79, at Vesuvius; yet these objects are no larger than many which, if they were situated on our satellite, would be termed comparatively small, if not insignificant.

One of the principal aims of lunar research is to learn as much as possible as to the present condition of the surface. Every one qualified to give an opinion will admit that this cannot be accomplished by roaming at large over the whole visible superficies, but only by confining attention to selected areas of limited extent, and recording and describing every object visible thereon, under various conditions of illumination, with the greatest accuracy attainable. This plan was suggested and inaugurated nearly thirty years ago by Mr. Birt, under the patronage of the British Association; but as he proposed to deal with the entire disc in this way, the magnitude and ambitious character of the scheme soon damped the ardour of those who at first supported it, and it was ultimately abandoned. It was, however, based on the only feasible principle which, as it seems to the writer, will not result in doubt and confusion. Now that photography has come to the assistance of the observer, Mr. Birt's proposal, if confined within narrower limits, would be far less arduous an undertaking than before, and might be easily carried out. A complete photographic survey of a few selected regions, as a basis for an equally thorough and exhaustive scrutiny by direct observation, would, it is believed, lead to a much more satisfactory and hopeful method for ultimately furnishing irrefragable testimony as to permanency or change than any that has yet been undertaken.

CATALOGUE OF LUNAR FORMATIONS

FIRST QUADRANT

WEST LONGITUDE 90 deg. TO 60 deg.

SCHUBERT.--This ring-plain, about 46 miles in diameter, situated on the N.E. side of the Mare Smythii, is too near the limb to be well observed.

NEPER.--Though still nearer the limb, this walled-plain, 74 miles in diameter, is a much more conspicuous object. It has a lofty border and a prominent central mountain, the highest portion of a range of hills which traverses the interior from N. to S.

APOLLONIUS.--A ring-plain, 30 miles in diameter, standing in the mountainous region S. of the Mare Crisium. There is a large crater on the S.W.

wall, and another, somewhat smaller, adjoining it on the N. There are many brilliant craters in the vicinity.

FIRMICUS.--A somewhat larger, more regular, but, in other respects, very similar ring-plain, N.W. of the last. Some distance on the W., Madler noted a number of dark-grey streaks which apparently undergo periodical changes, suggestive of something akin to vegetation. They are situated near a prominent mountain situated in a level region.

AZOUT.--A small ring-plain, connected with the last by a lofty ridge. It is the apparent centre of many other ridges and valleys which radiate from it towards the N.W. and the Mare Crisium. There is a central mountain, not an easy telescopic object, on its dusky floor.

CONDORCET.--A very prominent ring-plain, 45 miles in diameter, situated on the mountainous S.W. margin of the Mare Crisium. It is encircled by a lofty wall about 8000 feet in height. The dark interior of this and of the three preceding formations render them easily traceable under a high angle of illumination.

HANSEN.--A ring-plain, 32 miles in diameter, on the W. border of the Mare Crisium N. of Condorcet. Schmidt shows a central mountain and a terraced wall.

ALHAZEN.--This ring-plain, rather smaller than the last, is the most northerly of the linear chain of formations, associated with the highlands bordering the S.W. and the W. flanks of the Mare Crisium. It has a central mountain and other minor elevations on the floor. There is a little ring between Alhazen and Hansen, never very conspicuous in the telescope, which is plainly traceable in good photographs.

EIMMART.--A conspicuous ring-plain with bright walls on the N.W. margin of the Mare Crisium. The E. border attains a height of 10,000 feet above the interior, which, according to Schmidt, has a small central mountain. There is a rill-like valley on the E. of the formation.

ORIANI.--An irregular object, 32 miles in diameter, somewhat difficult to identify, N.W. of the last. There is a conspicuous crater on the N. of it, with which it is connected by a prominent ridge.

PLUTARCH.--A fine ring-plain W. of Oriani, with regular walls, and, according to Neison, with two central mountains, only one of which I have seen. Both this formation and the last are beautifully shown in a photograph taken August 19, 1891, at the Lick Observatory, when the moon's age was 15 d. 10 hrs.

SENECA.--Rather smaller than Plutarch. Too near the limb for satisfactory observation. Schmidt shows two considerable mountains in the interior. The position of this object in Schmidt's chart is not accordant with its place in Beer and Madler's map, nor in that of Neison.

HAHN.--A ring-plain, 46 miles in diameter, with a fine central mountain and lofty peaks on the border, which is not continuous on the S. There is a large and prominent crater on the E.

BEROSUS.--A somewhat smaller object of a similar type, N. of Hahn, but with a loftier wall. There is a want of continuity also in the border, the eastern and western sections of which, instead of joining, extend for some distance towards the S., forming a narrow gorge or valley. Outside the S.E. wall there is a small crater, and some irregular depressions on the E. The minute central mountain is only seen with difficulty under a low evening sun. The bright region between Hahn and Berosus and the western flank of Cleomedes is an extensive plain, devoid of prominent detail, and which, according to Neison, includes an area of 40,000 square miles.

GAUSS.--A large, and nearly circular walled-plain, 111 miles in diameter, situated close to the N.W. limb, and consequently always foreshortened into a more or less elongated ellipse. But for this it would be one of the grandest objects in the first quadrant. Under the designation of "Mercurius Falsus" it received great attention from Schroter, who gives several representations of it in his Selenotopographische Fragmente, which, though drawn in his usual conventional style, convey a juster idea of its salient features than many subsequent drawings made under far better optical conditions. The border, especially on the W., is very complex, and is discontinuous on the S., where it is intersected by more than one pass, and is prolonged far beyond the apparent limits of the formation. The most noteworthy feature is the magnificent mountain chain which traverses the floor from N. to S. It is interesting to watch the progress of sunset thereon, and see peak after peak disappear, till only the great central boss and a few minute glittering points of light, representing the loftier portions of the chain, remain to indicate its position. Madler expatiates on the sublime view which would be obtained by any one standing on the highest peak and observing the setting sun on one side of him and the nearly "full" earth on the other; while beneath him would lie a vast plain, shrouded in darkness, surrounded by the brilliantly illuminated peaks on the lofty border, gradually passing out of sunlight. In addition to the central mountain range, there are some large rings, craters, hillocks, &c., on the floor; and on the inner slope of the W. border there is a very large circular enclosure resembling a ring-plain, not recorded in the maps. Schmidt shows a row of large craters on the outer slope of the E. border. Of these, one is very conspicuous under a low evening sun, by reason of its brilliant walls and interior. In the region between Gauss and Berosus is a number of narrow steep ridges which follow the curvature of the E. wall.

STRUVE.--A small irregularly-shaped formation, open towards the S., forming one of the curious group of unsymmetrical enclosures associated with Messala. Its dark floor and a small dusky area on the N. indicate its position under a high sun.

CARRINGTON.--A small ring-plain, belonging to the Messala group, adjoining Schumacher on the N.W.

MERCURIUS.--This formation is 25 miles in diameter. A small crater stands on the S.E. section of the wall. There is a longitudinal range in the interior, and on the W. and N.W. the remains of two large walled-plains, the more westerly of which is a noteworthy object under suitable conditions. A short distance S. is a large, irregular, and very dark marking. On the N., lies an immense bright plain, extending nearly to the border of Endymion.

WEST LONGITUDE 60 deg. TO 40 deg.

TARUNTIUS.--Notwithstanding its comparatively low walls, this ring-plain, 44 miles in diameter, is a very conspicuous object under a rising sun. Like Vitello and a few other formations, it has an inner ring on the floor, concentric with the outer rampart, which I have often seen nearly complete under evening illumination. There is a small bright crater on the S.E. wall, and a larger one on the crest of the N.E. wall, with a much more minute depression on the W. of it, the intervening space exhibiting signs of disturbance. The upper portion of the wall is very steep, contrasting in this respect with the very gentle inclination of the glacis, which on the S. extends to a distance of at least 30 miles before it sinks to the level of the surrounding country, the gradient probably being as slight as 1 in 45. Two low dusky rings and a long narrow valley with brilliant flanks are prominent objects on the plain E. of Taruntius under a low evening sun.

SECCHI.--A partially enclosed little ring-plain S. of Taruntius, with a prominent central mountain and bright walls. There is a short cleft running in a N.E. direction from a point near the E. wall. Schmidt represents it as a row of inosculating craters.

PICARD.--The largest of the craters on the surface of the Mare Crisium, 21 miles in diameter. The floor, which includes a central mountain, is depressed about 2000 feet below the outer surface, and is surrounded by walls rising some 3000 feet above the Mare. A small but lofty ring-plain, Picard E, on the E., near the border of the Mare, is remarkable for its change of aspect under different angles of illumination. A long curved ridge running S. from this, with a lower ridge on the west, sometimes resemble a large enclosure with a central mountain. Still farther S., there is another bright deep crater, a, with a large low ring adjoining it on the S., abutting on the S.E. border of the Mare. Schroter bestowed much attention on these and other formations on the Mare Crisium, and attributed certain changes which he observed to a lunar atmosphere.

PEIRCE.--This formation, smaller than Picard, is also prominent, its border being very bright. There is a central peak, which, though not an easy object, I once glimpsed with a 4 inch Cook achromatic, and have seen it two or three times since with an 8 1/2 inch Calver reflector. A small crater, detected by Schmidt, which I once saw very distinctly under evening illumination, stands on the floor at the foot of the W. wall. Peirce A, a deeper formation, lies a little N. of Peirce, and has also, according to Neison, a very slight central hill, which is only just perceptible under the most favourable conditions. Schmidt appears to have overlooked it.

PROCLUS.--One of the most brilliant objects on the moon's visible surface, and hence extremely difficult to observe satisfactorily. It is about 18 miles in diameter, with very steep walls, and, according to Schmidt, has a small crater on its east border, where Madler shows a break. It is questionable whether there is a central mountain. It is the centre of a number of radiating light streaks which partly traverse the Mare Crisium, and with those emanating from Picard, Peirce, and other objects thereon, form a very complicated system.

MACROBIUS.--This, with a companion ring on the W., is a very beautiful object under a low sun. It is 42 miles in diameter, and is encircled by a bright, regular, but complex border, some 13,000 feet in height above the floor. Its crest is broken on the E. by a large brilliant crater, and its continuity is interrupted on the N. by a formation resembling a large double crater, which is associated with a number of low rounded banks and ridges extending some distance towards the N.W., and breaking the continuity of the glacis. The W. wall is much terraced, and on the N.W. includes a row of prominent depressions, well seen when the interior is about half illuminated under a rising sun. The central mountain is of the compound type, but not at all prominent. The companion ring, Macrobius C, is terraced internally on the W., and the continuity of its N. border broken by two depressions. There is a rill-valley between its N.E. side and Macrobius.

CLEOMEDES.--A large oblong enclosure, 78 miles in diameter, with massive walls, varying in altitude from 8000 to 10,000 feet above the interior. The most noteworthy features in connection with the circumvallation are the prominent depressions on the W. wall. Under a rising sun, when about one-fourth of the floor is in shadow, three of these can be easily distinguished, each resembling in form the analemma figure. There are two other curious depressions at the S.W. end of the formation. On the dark steel-grey floor are two irregular dusky areas, and a narrow but bright central mountain, on which, according to Schmidt, stand two little craters. There are two ring-plains on the S.W. quarter, and a group of three associated craters on the N. side, one of which (A) Schroter believed came into existence after he commenced to observe the formation, a supposition that has been shown by Birt and others to be very improbable.

TRALLES.--A large irregular crater, one of the deepest on the visible surface of the moon, situated on the N.E. wall of Cleomedes. There is a crater on its N. wall, and, according to Schmidt, some ridges and three closely associated craters on the floor.

BURCKHARDT.--This object, situated on an apparent extension of the W. wall of Cleomedes, is 35 miles in diameter, with a lofty border, rising on the E. to an altitude of nearly 13,000 feet. It has a prominent central mountain and some low ridges on the floor, which, together with two minute craters on the S.W. wall, I have seen under a low angle of morning illumination. It is flanked both on the E. and W. by deep irregular depressions, which present the appearance of having once been complete formations.

GEMINUS.--A fine regular ring-plain, 54 miles in diameter, nearly circular, with bright walls, rising on the E. to a height of more than 12,000 feet, and on the opposite side to nearly 16,000 feet above the floor. Their crest is everywhere very steep, and the inner slope is much terraced. There is a small but conspicuous mountain in the interior; N. of which I have seen a long ridge, where Schmidt shows some hillocks. Two fine clefts are easily visible within the ring, one running for some distance on the S.E. side of the floor, mounting the inner slope of the S.W. border to the summit ridge (where it is apparently interrupted), and then striking across the plain in a S.W. direction. Here it is accompanied for a short distance by a somewhat coarser companion, running parallel to it on the N. The other cleft occupies a very similar position on the N.W. side of the floor at the inner foot of the wall. On several occasions, when observing this formation and the vicinity, I have been struck by its peculiar colour under a low evening sun. At this time the whole region appears to be of a warm light brown or sepia tone.

BERNOUILLI.--A very deep ring-plain on the W. side of Geminus. Under evening illumination its lofty W. wall, which rises to a height of nearly 13,000 feet above the floor, is conspicuously brilliant. This formation exhibits a marked departure from the circular type, being bounded by rectilineal sides. The inner slope of the W. wall is slightly terraced. The border on the S. is much lower than elsewhere, as is evident when the formation is on the evening terminator. On the N. is the deep crater Messala a.

NEWCOMB.--The most prominent of a group of formations standing in the midst of the Haemus Mountains. Its crest is nearly 12,000 feet above the floor, on which there are some hills.

MESSALA.--This fine walled-plain, nearly 70 miles in diameter, is, with its surroundings, an especially interesting object when observed under a low angle of illumination. Its complex border, though roughly circular, displays many irregularities in outline, due mainly to rows of depressions. The best view of it is obtained when the W. wall is on the evening terminator. At this phase, if libration is favourable, the manifold details of its very uneven and apparently convex floor are best seen. On the S.W. side is a group of large craters associated with a number of low hills, of which Schmidt shows five; but I have seen many more, together with several ridges between them and the E. wall. I noted also a cleft, or it may be a narrow valley, running from the foot of the N.W. wall towards the centre. On the floor, abutting on the N.E. border, is a semicircular ridge of considerable height, and beyond the border on the N.E. there is another curved ridge, completing the circle, the wall forming the diameter. This formation is clearly of more ancient date than Messala, as the N.E. wall of the latter has cut through it. Where Messala joins Schumacher there is a break in the border, occupied by three deep depressions.

SCHUMACHER.--A large irregular ring-plain, 28 miles in diameter, associated with the N. wall of Messala, and having other smaller rings adjoining it on the E. and N. The interior seems to be devoid of detail.

HOOKE.--Another irregular ring-plain, 28 miles in diameter, on the N.E. of Messala. There is a bright crater of considerable size on the S.W., which is said to be more than 6000 feet in depth, and, according to Neison, is visible as a white spot at full. There is a smaller crater on the slope of the N.W. wall.

SHUCKBURGH.--A square-shaped enclosure on the N. of the last, with a comparatively low border. It has a conspicuous crater at its N.W. corner.

BERZELIUS.--A considerable ring-plain of regular form, with low walls and dark interior, on which there is a central peak, difficult to detect.

FRANKLIN.--A ring-plain, 33 miles in diameter, which displays a considerable departure from the circular type, as the border is in great part made up of rectilineal sections. Both the W. and N.E. wall is much terraced, and rises about 8000 feet above the dark floor, on the S. part of which there is a long ridge. There is a bright little isolated mountain on the plain E. of the formation, and a conspicuous craterlet on the N.W. An incomplete ring, with a very attenuated border, abuts on the S. side of Franklin.

CEPHEUS.--A peculiarly shaped ring-plain, 27 miles in diameter. The E. border is nearly rectilineal, while on the W., the wall forms a bold curve. There is a very brilliant crater on the summit of this section, and a central mountain on the floor. The W. wall is much terraced. W. of Cepheus, close to the brilliant crater, there is a cleft or narrow valley running N. towards Oersted.

OERSTED.--An oblong formation with very low walls, scarcely traceable on the S.E., except when near the terminator. There is a conspicuous crater on the N.W. side of the floor, and a curious square enclosure, with a crater on its W. border, abutting on the N.E. wall.

CHEVALLIER.--An inconspicuous object enclosed by slightly curved ridges. It includes a deep bright crater. On the N. is a low square formation and a long ridge running N. from it. Just beyond the N.E. wall is the fine large crater, Atlas A, with a much smaller but equally conspicuous crater beyond. A has a central hill, which, in spite of the bright interior, is not a difficult feature.

ATLAS.--This, and its companion Hercules on the E., form under oblique illumination a very beautiful pair, scarcely surpassed by any other similar objects on the first quadrant. Its lofty rampart, 55 miles in diameter, is surmounted by peaks, which on the N. tower to an altitude of nearly 11,000 feet. It exhibits an approach to a polygonal outline, the lineal character of the border being especially well marked on the N. The detail on the somewhat dark interior will repay careful scrutiny with high powers. There is a small but distinct central mountain, south of which stands a number of smaller hills, forming with the first a circular arrangement, suggestive of the idea that they represent the relics of a large central crater. Several clefts may be seen on the floor under suitable illumination, among them a forked cleft on the N.E. quarter, and two others, originating at a dusky pit of irregular form situated near the foot of the S.E. wall, one of which runs W. of the central hills, and the other on the opposite side. A ridge, at times resembling a light marking, extends from the central mountain to the N. border. During the years 1870 and 1871 I bestowed some attention on the dusky pit, and was led to suspect that both it and the surrounding area vary considerably in tone from time to time. Professor W.H. Pickering, observing the formation in 1891 with a 13 inch telescope under the favourable atmospheric conditions which prevail at Arequipa, Peru, confirmed this supposition, and has discovered some very interesting and suggestive facts relating to these variations, which, it is hoped, will soon be made public. On the plain a short distance beyond the foot of the glacis of the S.E. wall, I have frequently noted a second dusky spot, from which proceeds, towards the E., a long rill-like marking. On the N. there is a large formation enclosed by rectilineal ridges. The outer slopes of the rampart of Atlas are very noteworthy under a low sun.

HERCULES.--The eastern companion of Atlas, a fine ring-plain, about 46 miles in diameter, with a complex border, rising some 11,000 feet above a depressed floor. There are few formations of its class and size which display so much detail in the shape of terraces, apparent landslips, and variation in brightness. In the interior, S.E. of the centre, is a very conspicuous crater, which is visible as a bright spot when the formation itself is hardly traceable, two large craterlets slightly N. of the centre, and several faint little spots on the east of them. The latter, detected some years ago by Herr Hackel of Stuttgart, are arranged in the form of a horse-shoe. There are two small contiguous craters on the S.E. wall, one of which, a difficult object, was recently detected by Mr. W.H. Maw, F.R.A.S. The well-known wedge-shaped protuberance on the S. wall is due to a large irregular depression. On the bright inner slope of the N. wall are manifest indications of a landslip.

ENDYMION.--A large walled-plain, 78 miles in diameter, enclosed by a lofty, broad, bright border, surmounted in places by peaks which attain a height of more than 10,000 feet above the interior, one on the W. measuring more than 15,000 feet. The walls are much terraced and exhibit two or three breaks. The dark floor appears to be devoid of detail. Schmidt, however, draws two large irregular mounds E. of the centre, and shows four narrow light streaks crossing the interior nearly parallel to the longer axis of the formation.

DE LA RUE.--Notwithstanding its great extent, this formation hardly deserves a distinctive name, as from the lowness of its border it is scarcely traceable in its entirety except under very oblique light. Schmidt, nevertheless, draws it with very definite walls, and shows several ridges and small rings in the interior. Among these objects, a little E. of the centre, there is a prominent peak.

STRABO.--A small walled-plain, 32 miles in diameter, connected with the N. border of the last.

THALES.--A bright formation, also associated with the N. side of De la Rue, adjoining Strabo on the N.E. Schmidt shows a minute hill in the interior.

There are several unnamed formations, large and small, between De la Rue and the limb, some of which are well worthy of examination.

WEST LONGITUDE 40 deg. TO 20 deg.

MASKELYNE.--A regular ring-plain, 19 miles in diameter, standing almost isolated in the Mare Tranquilitatis. The floor, which includes a central mountain, is depressed some 3000 feet below the surrounding surface. There are prominent terraces on the inner slope of the walls. Schmidt shows no craters upon them, but Madler draws a small one on the E., the existence of which I can confirm.

MANNERS.--A brilliant little ring-plain, 11 miles in diameter, on the S.E. side of the Mare Tranquilitatis. There appears to be no detail whatever in connection with its wall. It has a distinct central mountain. About three diameters distant on the S.W. there is a bright crater, omitted by Madler and Neison.

ARAGO.--A much larger formation, 18 miles in diameter, N. of the last, with a small crater on its N. border, and exhibiting two or three spurs from the wall on the opposite side. The inner slopes are terraced, and there is a small central mountain. There are two curious circular protuberances on the Mare E. of Arago, which are well seen when the W. longitude of the morning terminator is about 19 deg., and a long cleft, passing about midway between them, and extending from the foot of the E. wall to a small crater on the edge of the Mare near Sosigenes. Another cleft, also terminating at this crater, runs towards Arago and the more northerly of the protuberances.

CAUCHY.--A bright little crater, not more than 7 or 8 miles in diameter, on the W. side of the Mare Tranquilitatis, N.E. of Taruntius. It has a peak on its W. rim considerably loftier than the rest of the wall, which is visible as a brilliant spot at sunrise long before the rest of the rampart is illuminated. On the S. there are two bright longitudinal ridges ranging from N.E. to S.W. These stand in the position where Neison draws two straight clefts. The Cauchy cleft, however, lies N. of these, and terminates, as shown by Schmidt, among the mountains N.E. of Taruntius. I have seen it thus on many occasions, and it is so represented in a drawing by M.E. Stuvaert (Dessins de la Lune). There is a number of minute craters and mounds standing on the S. side of this cleft, and many others in the vicinity.

JANSEN.--Owing to its comparatively low border, this is not a very conspicuous object. It is chiefly remarkable for the curious arrangement of the mountains and ridges on the S. and W. of it. There is a bright little crater on the S. side of the floor, and many noteworthy objects of the same class in the neighbourhood. The mountain arm running S., and ultimately bending E., forms a large incomplete hook-shaped formation terminating at a ring-plain, Jansen B. The ridges in the Mare Tranquilitatis between Jansen B. and the region E. of Maskelyne display under a low sun foldings and wrinklings of a very extraordinary kind.

MACLEAR.--A conspicuous ring-plain about 16 miles in diameter. The dark floor includes, according to Madler, a delicate central hill which Schmidt does not show. Neison, however, saw a faint greyish mark, and an undoubted peak has been subsequently recorded. I have not succeeded in seeing any detail within the border, which in shape resembles a triangle with curved sides.

ROSS.--A somewhat larger ring-plain of irregular form, on the N.W. of the last. There are gaps on the bright S.W. border and a crater on the S.E. wall. The central mountain is an easy feature.

PLINIUS.--This magnificent object reminds one at sunrise of a great fortress or redoubt erected to command the passage between the Mare Tranquilitatis and the Mare Serenitatis. It is 32 miles in diameter, and is encompassed by a very massive rampart, rising at one peak on the E. to more than 6000 feet above the interior, and displaying, especially on the S.E., and N., many spurs and buttresses. The exterior slopes at sunrise, and even when the sun is more than 10 deg. above the horizon, are seen to be traversed by wide and deep valleys. The S. glacis is especially broad, extending to a distance of 10 or 12 miles before it runs down to the level of the plain. The shape of the circumvallation, when it is fully illuminated, approximates very closely to that of an equilateral triangle with curved sides. There are two bright little craters on the outer slope, just below the summit ridge on the S.E., and another, larger, on the N. wall, in which it makes a prominent gap. The interior is considerably brighter than the surface of the surrounding Mare, and, a little S. of the centre, includes two crater-like objects with broken rims. These assume different aspects under different conditions of illumination, and it is only when the floor is lighted by a comparatively low morning sun, that their true character is apparent. On the N.W. quarter of the interior are two smaller distinct craters, and a square arrangement of ridges. On the N.E. there are some hillocks and minor elevations. The Plinius rills form an especially interesting system, and under favourable conditions may be seen in their entirety with a good 4 inch refractor, about the time when the morning terminator passes through Julius Caesar. They consist of three long fissures, originating amid the Haemus highlands, on the S. side of the Mare Serenitatis, and diverging towards the W. The most southerly commences S.S.E. of the Acherusian promontory (a great headland, 5000 feet high, at the W. termination of the Haemus range), and, following a somewhat undulating course, runs up to the N. side of Dawes. Under a low evening sun, I have remarked many inequalities in the width of that portion of it immediately N. of Plinius, which appear to indicate that it is here made up of rows of inosculating craters. The cleft north of this originates very near it, passes a little S. of the promontory, and runs to the E. edge of the plateau surrounding Dawes. The third and most northerly cleft begins at a point immediately N. of the promontory, cuts through the S. end of the well-known Serpentine ridge on the Mare Serenitatis, and, after following a course slightly concave to the N., dies out on the N. side of the plateau. This cleft forms the line of demarcation between the dark tone of the Mare Serenitatis and the light hue of the Mare Tranquilitatis, traceable under nearly every condition of illumination, and prominent in all good photographs.

DAWES.--A ring-plain 14 miles in diameter, situated N.W. of Plinius, on a nearly circular light area. Its bright border rises to a height of 2000 feet above the Mare, and includes a central mountain, a white marking on the E., and a ridge running from the mountain to the S. wall. There are two closely parallel clefts on the N. side of the plateau running from E. to W., that nearer Dawes being the longer, and having a craterlet standing upon it about midway between its extremities. At its W. termination there is a crater-row running at right angles to it. The light area appears to be bounded on the E. by a low curved bank.

VITRUVIUS.--A ring-plain 19 miles in diameter with bright but not very lofty walls, situated among the mountains near the S.W. side of the Mare Serenitatis. It is surrounded by a region remarkable for its great variability in brightness. There is a large bright ring-plain on the W., with a less conspicuous companion on the S. of it.

MARALDI.--A deep but rather inconspicuous formation, bounded on the W. by a polygonal border. A small ring-plain with a central mountain is connected with the S.W. wall; and, running in a N. direction from this, is a short mountain arm which joins a large circular enclosure with a low broken border standing on the N. side of the Mare Tranquilitatis.

LITTROW.--A peculiar ring-plain, rather smaller than the last, some distance N. of Vitruvius, on the rocky W. border of the Mare Serenitatis. It is shaped like the letter D, the straight side facing the W. There is a distinct crater on the N. wall. On the N.W. it is flanked by three irregular ring-plains, and on the S.E. by a fourth. Neison shows two small mountains on the floor, but Schmidt, whose drawing is very true to nature, has no detail whatever. A fine cleft may be traced from near the foot of the E. wall to Mount Argaeus, passing S. of a bright crater on the Mare E. of Littrow. It extends towards the Plinius system, and is probably connected with it.

MOUNT ARGAEUS.--There are few objects on the moon's visible surface which afford a more striking and beautiful picture than this mountain and its surrounding heights with their shadows a few hours after sunrise. It attains an altitude of more than 8000 feet above the Mare, and at a certain phase resembles a bright spear-head or dagger. There is a well- defined rimmed depression abutting on its southern point.

ROMER.--A prominent formation of irregular outline, 24 miles in diameter, situated in the midst of the Taurus highlands. It has a very large central mountain, a crater on the N. side of the floor, and terraced inner slopes.

Some distance on the N. is another ring, nearly as large, with a crater on its S. rim, and between this and Posidonius is another with a wide gap on the S. and a crater on its N. border. One of the most remarkable crater-rills on the moon runs from the E. side of Romer through this latter ring, and then northwards on to the plain W. of Posidonius. Under suitable conditions, it can be seen as such in a 4 inch achromatic. It is easily traceable as a rill in a photograph of the N. polar region of the moon taken by MM. Henry at the Paris Observatory, and recently published in Knowledge.

LE MONNIER.--A great inflection or bay on the W. border of the Mare Serenitatis S. of Posidonius. Like many other similar formations on the edges of the Maria, it appears at one time or other to have had a continuous rampart, which on the side facing the "sea" has been destroyed. In this, as in most of the other cases, relics of the ruin are traceable under oblique light. A fine crescent-shaped mountain, 3000 feet high, stands near the S. side of the gap, and probably represents a portion of a once lofty wall. It will repay the observer to watch the progress of sunrise on the whole of the W. coast-line of the Mare up to Mount Argaeus.

POSIDONIUS.--This magnificent ring-plain is justly regarded as one of the finest telescopic objects in the first quadrant. Its narrow bright wall with its serrated shadow, the conspicuous crater, the clefts and ridges and other details on the floor, together with the beautiful group of objects on the neighbouring plain, and the great Serpentine ridge on the E., never fail to excite the interest of the observer. The circumvallation, which is far from being perfectly regular, is about 62 miles in diameter, and, considering its size, is not remarkable for its altitude, as it nowhere exceeds 6000 feet above the interior, which is depressed about 2000 feet below the surrounding plain. Its continuity, especially on the E., is interrupted by gaps. On the N., the wall is notably deformed. It is broader and more regular on the W., where it includes a large longitudinal depression, and on the N.W. section stand two bright little ring-plains. On the floor, which shines with a glittering lustre, are the well-marked remains of a second ring, nearly concentric with the principal rampart, and separated from it by an interval of nine or ten miles. The most prominent object, however, is the bright crater a little E. of the centre. This is partially surrounded on the W. by three or four small bright mountains, through which runs in a meridional direction a rill-valley, not easily traced as a whole, except under a low sun. There is another cleft on the N.E. side of the interior, which is an apparent extension of part of the inner ring, a transverse rill-valley on the N., a fourth quasi rill on the N.W., and a fifth short cleft on the S. part of the floor. Between the principal crater and the S.E. wall are two smaller craters, which are easy objects. Beyond the border on the N., in addition to Daniell, are four conspicuous craters and many ridges.

CHACORNAC.--This object, connected with Posidonius on the S.W., is remarkable for the brilliancy of its border and the peculiarity of its shape, which is very clearly that of an irregular pentagon with linear sides. I always find the detail within very difficult to make out. Two or more low ridges, traversing the floor from N. to S., and a small crater, are, however, clearly visible under oblique illumination. Schmidt draws a crater-rill, and Neison two parallel rills on the floor,--the former extends in a southerly direction to the W. side of Le Monnier.

DANIELL.--A bright little ring-plain N. of Posidonius. It is connected with a smaller ring-plain on the N.W. wall of the latter by a low ridge.

BOND, G.P.--A small bright ring-plain 12 miles in diameter, W. of Posidonius. Neison shows a crater both on the N. and S. rim. Schmidt omits these.

MAURY.--A bright deep little ring-plain, about 12 miles in diameter, on the W. border of the Lacus Somniorum. It is the centre of four prominent hill ranges.

GROVE.--A bright deep ring-plain, 15 miles in diameter, in the Lacus Somniorum, with a border rising 7000 feet above a greatly depressed floor, which includes a prominent mountain.

MASON.--The more westerly of two remarkable ring-plains, situated in the highlands on the S. side of the Lacus Mortis. It is 14 miles in diameter, has a distinct crater on its S. wall, and, according to Schmidt, a crater on the E. side of the floor.

PLANA.--A formation 23 miles in diameter, closely associated with the last. Neison states that the floor is convex and higher than the surrounding region. It has a triangular-shaped central mountain, a crater, and at least three other depressions on the S.W. wall where it joins Mason.

BURG.--A noteworthy formation, 28 miles in diameter, on the Mare, N. of Plana. The floor is concave, and includes a very large bright mountain, which occupies a great portion of it. The interior slopes are prominently terraced, and there are several spurs associated with the glacis on the S. and N.E. A distinct cleft runs from the N. side of the formation to the S.E. border of the Lacus Somniorum, which is crossed by another winding cleft running from a crater E. of Plana towards the N.E.

BAILY.--A small ring-plain, N. of Burg, flanked by mountains, with a large bright crater on the W. The group of mountains standing about midway between it and Burg are very noteworthy.

GARTNER.--A very large walled-plain with a low incomplete border on the E., but defined on the W. by a lofty wall. Schmidt shows a curved crater-row on the W. side of the floor.

DEMOCRITUS.--A deep regular ring-plain, about 25 miles in diameter, with a bright central mountain and lofty terraced walls.

ARNOLD.--A great enclosure, bounded, like so many other formations hereabouts, by straight parallel walls. There is a somewhat smaller walled-plain adjoining it on the W.

MOIGNO.--A ring-plain with a dark floor, adjoining the last on the N.E. There is a conspicuous little crater in the interior.

EUCTEMON.--This object is so close to the limb that very little can be made of its details under the most favourable conditions. According to Neison, there is a peak on the N. wall 11,000 feet in height.

METON.--A peculiarly-shaped walled-plain of great size, exhibiting considerable parallelism. The floor is seen to be very rugged under oblique illumination.

WEST LONGITUDE 20 deg. TO 0 deg.

SABINE.--The more westerly of a remarkable pair of ring-plains, of which

Ritter is the other member, situated on the E. side of the Mare Tranquilitatis a little N. of the lunar equator. It is about 18 miles in diameter, and has a low continuous border, which includes a central mountain on a bright floor. From a mountain arm extending from the S. wall, run in a westerly direction two nearly parallel clefts skirting the edge of the Mare. The more southerly of these terminates near a depression on a rocky headland projecting from the coast-line, and the other stops a few miles short of this. A third cleft, commencing at a point N.E. of the headland, runs in the same direction up to a small crater near the N. end of another cape-like projection. At 8 h. on April 9, 1886, when the morning terminator bisected Sabine, I traced it still farther in the same direction. All these clefts exhibit considerable variations in width, but become narrower as they proceed westwards.

RITTER.--Is very similar in every respect to the last. A curved rill mentioned by Neison is on the N.E. side of the floor and is concentric with the wall. On the N. side of this ring-plain are three conspicuous craters, the two nearer being equal in size and the third much smaller.

SCHMIDT.--A bright crater at the foot of the S. slope of Ritter.

DIONYSIUS.--This crater, 13 miles in diameter, is one of the brightest spots on the lunar surface. It stands on the E. border of the Mare, about 30 miles E.N.E. of Ritter. A distinct crater-row runs round its outer border on the W., and ultimately, as a delicate cleft, strikes across the Mare to the E. side of Ritter. Both crater-row and cleft are easy objects in a 4 inch achromatic under morning illumination.

ARIADAEUS.--A bright little crater of polygonal shape, with another crater of about one-third the area adjoining it on the N.W., situated on the rocky E. margin of the Mare Tranquilitatis, N.E. of Ritter. A short cleft runs from it towards the latter, but dies out about midway. A second cleft begins near its termination, and runs up to the N.E. wall of Ritter. E. of this pair a third distinct cleft, originating at a point on the coast-line about midway between Ariadaeus and Dionysius, ends near the same place on the border. There is a fourth cleft extending from the N. side of a little bay N. of Ariadaeus across the Mare to a point N.W. of the more northerly of the three craters N. of Ritter. At a small crater on the S. flank of the mountains bordering the little bay N. of Ariadaeus originates one of the longest and most noteworthy clefts on the moon's visible surface, discovered more than a century ago by Schroter of Lilienthal. It varies considerably in breadth and depth, but throughout its course over the plain, between Ariadaeus and Silberschlag, it can be followed without difficulty in a very small telescope. E. of the latter formation, towards Hyginus (with which rill-system it is connected), it is generally more difficult. A few miles E. of Ariadaeus it sends out a short branch, running in a S.W. direction, which can be traced as a fine white line under a moderately high sun. It is interesting to follow the course of the principal cleft across the plain, and to note its progress through the ridges and mountain groups it encounters. In the great Lick telescope it is seen to traverse some old crater-rings which have not been revealed in smaller instruments. About midway between Ariadaeus and Silberschlag it exhibits a duplication for a short distance, first detected by Webb.

DE MORGAN.--A brilliant little crater, 4 miles in diameter, on the plain S. of the Ariadaeus cleft.

CAYLEY.--A very deep bright crater, with a dark interior, N. of the last, and more than double its diameter. There is a second crater between this and the cleft.

WHEWELL.--Another bright little ring, about 3 miles in diameter, some distance to the E. of De Morgan and Cayley.

SOSIGENES.--A small circular ring-plain, 14 miles in diameter, with narrow walls, a central mountain, and a minute crater outside the wall on the E.; situated on the E. side of the Mare Tranquilitatis, W. of Julius Caesar. There is another crater, about half its diameter, on the S., connected with it by a low mound. This has a still smaller crater on the W. of it.

JULIUS CAESAR.--A large incomplete formation of irregular shape. The wall on the E. is much terraced, and forms a flat "S" curve. The summit ridge is especially bright, and has a conspicuous little crater upon it. On the W. is a number of narrow longitudinal valleys trending from N. to S., included by a wide valley which constitutes the boundary on this side. The border on the S. consists of a number of low rounded banks, those immediately E. of Sosigenes being traversed by several shallow valleys, which look as if they had been shaped by alluvial action. There is a brilliant little hill at the end of one of these valleys, a few miles E. of Sosigenes. The floor of Julius Caesar is uneven in tone, becoming gradually duskier from S. to N., the northern end ranking among the darkest areas on the lunar surface. There are at least three large circular swellings in the interior. A long low mound, with two or three depressions upon it, bounds the wide valley on the E. side.

GODIN.--A square-shaped ring-plain, 28 miles in diameter, with rounded corners. The bright rampart is everywhere lofty, except on the S., is much terraced, and includes a central mountain. On the S. a curious trumpet-shaped valley, extending some distance towards the S.W., and bounded by bright walls, is a noteworthy feature at sunrise. There are other longitudinal valleys with associated ridges on this side of the formation, all running in the same direction. There is a large bright crater outside the border on the N.E., and, between it and the wall, another, smaller, which is readily seen under a high sun.

AGRIPPA.--A ring-plain 28 miles in diameter on the N. of the last, with a terraced border rising to a height of between 7000 and 8000 feet above the floor, which contains a large bright central mountain and two craters on the S. The shape of this formation deviates very considerably from circularity, the N. wall, on which stands a small crater, being almost lineal. On the W., at a distance of a few miles, runs the prominent mountain range, extending northwards nearly up to the E. flank of Julius Caesar, which bounds the E. side of the great Ariadaeus plain. Between this rocky barrier and Agrippa is a very noteworthy enclosure containing much minute detail and a long straight ridge resembling a cleft. A few miles N. of Agrippa stands a small crater; at a point W. of which the Hyginus cleft originates.

SILBERSCHLAG.--A very brilliant crater, 8 or 9 miles in diameter, connected with the great mountain range just referred to. The Ariadaeus cleft cuts through the range a few miles N. of it. This neighbourhood at sunrise presents a grand spectacle. With high powers under good atmospheric conditions, the plain E. of the mountains is seen to be traversed by a number of shallow winding valleys, trending towards Agrippa, and separated by low rounded hills which have all the appearance of having been moulded by the action of water.

BOSCOVICH.--This is not a very striking telescopic object under any phase, on account of its broken, irregular, and generally ill-defined border. It is, however, remarkable as being one of the darkest spots on the visible surface: in this respect a fit companion to Julius Caesar, its neighbour on the W. Schmidt shows some ridges within it.

RHAETICUS.--A very interesting formation, about 25 miles in diameter, situated near the lunar equator, with a border intersected by many passes. A deep rill-like valley winds round its eastern glacis, commencing on the S. at a small circular enclosure standing at the end of a spur from the wall; and, after crossing a ridge W. of a bright little crater on the N. of the formation, apparently joins the most easterly cleft of the Triesnecker system. A cleft traverses the N. side of the floor of Rhaeticus, and extends across the plain on the E. as far as the N. side of Reaumur.

TRIESNECKER.--Apart from being the centre of one of the most remarkable rill-systems on the moon, this ring-plain, though only about 14 miles in diameter, is an object especially worthy of examination under every phase. At sunrise, and for some time afterwards, owing to the superior altitude of the N.W. section of the wall, a considerable portion of the border on the N. and N.E. is masked by its shadow, which thus appears to destroy its continuity. On more than one occasion, friends, to whom I have shown this object under these conditions, have likened it to a breached volcanic cone, a comparison which at a later stage is seen to be very inappropriate. The rampart is terraced within, and exhibits many spurs and buttresses without, especially on the N.W. The central mountain is small and not conspicuous. The rill-system is far too complicated to be intelligibly described in words. It lies on the W. side of the meridian passing through the formation, and extends from the N. side of Rhaeticus to the mountain-land lying between Ukert and Hyginus on the N. Birt likened these rills to "an inverted river system," a comparison which will commend itself to most observers who have seen them on a good night, for in many instances they appear to become wider and deeper as they approach higher ground. Published maps are all more or less defective in their representations of them, especially as regards that portion of the system lying N. of Triesnecker.

HYGINUS.--A deep depression, rather less than 4 miles across, with a low rim of varying altitude, having a crater on its N. edge. This formation is remarkable for the great cleft which traverses it, discovered by Schroter in

1788. The coarser parts of this object are easily visible in small telescopes, and may be glimpsed under suitable conditions with a 2 inch achromatic. Commencing a little W. of a small crater N. of Agrippa, it crosses, as a very delicate object, a plain abounding in low ridges and shallow valleys, and runs nearly parallel to the eastern extension of the Ariadaeus rill. As it approaches Hyginus it becomes gradually coarser, and exhibits many expansions and contractions, the former in many cases evidently representing craters. When the phase is favourable, it can be followed across the floor of Hyginus, and I have frequently seen the banks with which it appears to be bounded (at any rate within the formation), standing out as fine bright parallel lines amid the shadow. On reaching the E. wall, it turns somewhat more to the N., becomes still coarser and more irregular in breadth, and ultimately expands into a wide valley on the N.E. It is connected with the Ariadaeus cleft by a branch which leaves the latter at an acute angle on the plain E. of Silberschlag, and joins it about midway between its origin N. of Agrippa and Hyginus. It is also probably joined to the Triesnecker system by one or more branches E. of Hyginus.

On May 27, 1877, Dr. Hermann Klein of Cologne discovered, with a 5 1/2 inch Plosel dialyte telescope, a dark apparent depression without a rim in the Mare Vaporum, a few miles N.W. of Hyginus, which, from twelve years' acquaintance with the region, he was certain had not been visible during that period. On the announcement of this discovery in the Wochenschrift fur Astronomie in March of the following year, the existence of the object described by Dr. Klein was confirmed, and it was sedulously scrutinised under various solar altitudes. To most observers it appeared as an ill-defined object with a somewhat nebulous border, standing on an irregularly-shaped dusky area, with two or more small dark craters and many low ridges in its vicinity. A little E. of it stands a curious spiral mountain called the Schneckenberg. The question as to whether Hyginus N. (as the dusky spot is called) is a new object or not, cannot be definitely determined, as, in spite of a strong case in favour of it being so, there remains a residuum of doubt and uncertainty that can never be entirely cleared away. After weighing, however, all that can be said "for and against," the hypothesis of change seems to be the most probable.

UKERT.--This bright crater, 14 miles in diameter, situated in the region N.E. of Triesnecker, is surrounded by a very complicated arrangement of mountains; and on the N. and W. is flanked by other enclosures. It has a distinct central mountain. Its most noteworthy feature is the great valley, more than 80 miles long, which extends from N.E. to S.W. on the E. side of it. This gorge is at least six miles in breadth, of great depth, and is only comparable in magnitude with the well-known valley which cuts through the Alps, W. of Plato. A delicate cleft, not very clearly traceable as a whole, begins near its N. end, and terminates amid the ramifications of the Apennines S. of Marco Polo.

TAQUET.--A conspicuous little crater on the S. border of the Mare Serenitatis at the foot of the Haemus Mountains. A branch of the great Serpentine ridge, which traverses the W. side of this plain and other lesser elevations, runs towards it.

MENELAUS.--A conspicuously bright regular ring-plain, about 20 miles in diameter, situated on the S. coast-line of the Mare Serenitatis, and closely associated with the Haemus range. It has a brilliant central mountain, but no visible detail on the walls. On the edge of the Mare, S.W. of it, there is a curious square formation. The bright streak traversing the Mare from N. to S., which is so prominently displayed in old maps of the moon, passes through this formation.

SULPICIUS GALLUS.--Another brilliant object on the south edge of the Mare Serenitatis, some distance E. of the last. It is a deep circular crater about 8 miles in diameter, rising to a considerable height above the surface. Its shadow under a low morning sun is prominently jagged. On the E. are two bright mounds, and S. of that which is nearer the border of the Mare, commences a cleft which, following the curvature of the coast- line, terminates at a point in W. long. 9 deg. This object varies considerably in width and depth. Another shorter and coarser cleft runs S. of this across an irregularly shaped bay or inflexion in the border of the Mare.

MANILIUS.--This, one of the most brilliant objects in the first quadrant, is about 25 miles in diameter, with walls nearly 8000 feet above the floor, which includes a bright central mountain. The inner slope of the border on the E. is much terraced and contains some depressions. There is a small isolated bright mountain 2000 feet high on the Mare Vaporum, some distance to the E.

BESSEL.--A bright circular crater, 14 miles in diameter, on the S. half of the Mare Serenitatis, and the largest object of its class thereon. Its floor is depressed some 2000 feet below the surrounding surface, while the walls, rising nearly 1600 feet above the plain, have peaks both on the N. and S. about 200 feet higher. The shadows of these features, noted by Schroter in 1797, and by many subsequent observers, are very noteworthy. I have seen the shadow of a third peak about midway between the two. One may faintly imagine the magnificent prospect of the coast- line of the Mare with the Haemus range, which would be obtained were it possible to stand on the summit of one of these elevations. It is doubtful whether Bessel has a central mountain. Neither Madler nor Schmidt have seen one, though Webb noted a peak on two occasions. I fail to see anything within the crater. The bright streak crossing the Mare from N. to S. passes through Bessel.

LINNE.--A formation on the E. side of the Mare Serenitatis, described by Lohrmann and Madler as a deep crater, but which in 1866 was found by Schmidt to have lost all the appearance of one. The announcement of this apparent change led to a critical examination of the object by most of the leading observers, and to a controversy which, if it had no other result, tended to awaken an interest in selenography that has been maintained ever since. According to Madler, the crater was more than 6 miles in diameter in his time, and very conspicuous under a low sun, a description to which it certainly did not answer in 1867 or at any subsequent epoch. It is anything but an easy object to see well, as there is a want of definiteness about it under the best conditions, though the minute crater, the low ridges, and the nebulous whiteness described by Schmidt and noted by Webb and others, are traceable at the proper phase. As in the case of Hyginus N, there are still many sceptics as regards actual change, despite the records of Lohrmann and Madler; but the evidence in favour of it seems to preponderate.

CONON.--A bright little crater, 11 miles in diameter, situated among the intricacies of the Apennines, S. of Mount Bradley. It has a central hill, which is not a difficult object.

ARATUS.--One of the most brilliant objects on the visible surface of the moon, a crater 7 miles in diameter, S. of Mount Hadley, surrounded by the lofty mountain arms and towering heights of the Apennines. A peak close by on the N. is more than 10,000 feet, and another farther removed towards the

N.W. is over 14,000 feet in altitude.

AUTOLYCUS.--A ring-plain 23 miles in diameter, deviating considerably from circularity, W. of Archimedes, on the Mare Imbrium, or rather on that part of it termed the Palus Putredinis. Its floor, which contains an inconspicuous central mountain, is depressed some 4000 feet below the surrounding country. With a power of 150 on a 4 5/8 achromatic, Dr. Sheldon of Macclesfield has seen two shallow crateriform depressions in the interior, one nearly central, and the other about midway between it and the N. wall. The wall is terraced within, and has a crater just below its crest on the W., which, when the opposite border is on the morning terminator, is seen as a distinct notch. Autolycus is the centre of a minor ray-system.

ARISTILLUS.--A larger and much more elaborate ring-plain, 34 miles in diameter, N. of Autolycus. Its complex wall, with its terraces within, and its buttresses, radiating spurs, and gullies without, forms a grand telescopic object under a low sun on a good night. It rises on the east 11,000 feet above the Mare, and is about 2000 feet lower on the W., while the interior is depressed some 3000 feet. Its massive central mountain, surmounted by many peaks, occupies a considerable area on the floor, and exhibits a digitated outline at the base. On the S. and W. a number of deep valleys radiate from the foot of the border, some of them extending nearly as far as Autolycus. Shallower but more numerous and regular features of the same class radiate towards the N.E. from the foot of the opposite wall. On the N.W. are several curved ridges, all trending towards Theaetetus. On the S.E. the surface is trenched by a number of crossed gullies, well seen when the E. wall is on the morning terminator. Just beyond the N. glacis is a large irregular dusky enclosure with a central mound, and another smaller low ring adjoining it on the S.E. The visibility of these objects is very ephemeral, as they disappear soon after sunrise. Aristillus is also the centre of a bright ray system.

THEAETETUS.--A conspicuous ring-plain, about 16 miles in diameter, in the Palus Nebularum, N.W. of Aristillus. It is remarkable for its great depth, the floor sinking nearly 5000 feet below the surface. Its walls, 7000 feet high on the W., are devoid of detail. The glacis on the S.W. has a gentle slope, and extends for a great distance before it runs down to the level of the plain. Not far from the foot of the wall on the N. is a row of seven or eight bright little hills, near the eastern side of which originates a distinct cleft that crosses the

Palus in a N.W. direction, and terminates among mountains between Cassini and Calippus. I have seen this object easily with a 4 inch achromatic.

CALIPPUS.--A bright ring-plain 17 miles in diameter, situated in the midst of the intricate Caucasus Mountain range. On the E. is a brilliant peak rising more than 13,000 feet above the Palus Nebularum, and nearer the border, on the N.E., is a second, more than 500 feet higher, with many others nearly as lofty in the vicinity. Calippus has not apparently a central peak or any other features on the floor.

CASSINI.--This remarkable ring-plain, about 36 miles in diameter, is very similar in character to Posidonius. It has a very narrow wall, nowhere more than 4000 feet in height, and falling on the E. to 1500 feet. Though a prominent and beautiful object under a low sun, its attenuated border and the tone of the floor, which scarcely differs from that of the surrounding surface, render it difficult to trace under a high angle of illumination, and perhaps accounts for the fact that it escaped the notice of Hevel and Riccioli; though it is certainly strange that a formation which is thrown into such strong relief at sunrise and sunset should have been overlooked, while others hardly more prominent at these times have been drawn and described. The outline of Cassini is clearly polygonal, being made up of several rectilineal sections. The interior, nearly at the same level as the outside country, includes a large bright ring-plain, A, 9 miles in diameter and 2600 feet in depth, which has a good-sized crater on the S. edge of a great bank which extends from the S.W. side of this ring-plain to the wall. On the E. side of the floor, close to the inner foot of the border, is a bright deep crater about two-thirds of the diameter of A, and between it and the latter Brenner has seen three small hills. The outer slope of Cassini includes much detail. On the S.W. is a row of shallow depressions just below the crest of the wall, and near the foot of the slope is a large circular shallow depression associated with a valley which runs partly round it. The shape of the glacis on the W. is especially noteworthy, the S.W. and N.W. sides meeting at a slightly acute angle at a point 10 or 12 miles W. of the summit of the ring. On the outer E. slope is a curious elongated depression, and on the N. slope two large dusky rings, well shown by Schmidt, but omitted in other maps. Most of these details are well within the scope of moderate apertures. Perhaps the most striking view of Cassini and its surroundings is obtained when the morning terminator is on the central meridian.

ALEXANDER.--A large irregularly shaped plain, at least 60 miles in longest diameter, enclosed by the Caucasus Mountains. On the S.W. and N.W. the border is lineal. It has a dark level floor on which there is a great number of low hills.

EUDOXUS.--A bright deep ring-plain, about 40 miles in diameter, in the hilly region between the Mare Serenitatis and the Mare Frigoris, with a border much broken by passes, and deviating considerably from circularity. Its massive walls, rising more than 11,000 feet above the floor on the W., and about 10,000 feet on the opposite side, are prominently terraced, and include crater-rows in the intervening valleys, while their outer slopes present a complicated system of spurs and buttresses. There is a bright crater on the N. glacis, and some distance beyond the wall on the N.W. is a small ring-plain, and on the S.E. another, with a conspicuous crater between it and the wall. Neison draws attention to an area of about 1400 square miles on the N.E. which is covered with a great multitude of low hills. E. of Eudoxus are two short crossed clefts, and on the N. a long cleft of considerable delicacy running from N.E. to S.W. It was in connection with this formation that Trouvelot, on February 20, 1877, when the terminator passed through Aristillus and Alphonsus, saw a very narrow thread of light crossing the S. part of the interior and extending from border to border. He noted also similar appearances elsewhere, and termed them Murs enigmatiques.

ARISTOTELES.--A magnificent ring-plain, 60 miles in diameter, with a complex border, surmounted by peaks, rising to nearly 11,000 feet above the floor, one of which on the W., pertaining to a terrace, stands out as a brilliant spot in the midst of shadow when the interior is filled with shadow. The formation presents its most striking aspect at sunrise, when the shadow of the W. wall just covers the floor, and the brilliant inner slope of the E. wall with the little crater on its crest is fully illuminated. At this phase the details of the terraces are seen to the best advantage. The arrangement of the parallel ridges and rows of hills on the N.E. and S.W. is likewise better seen at this time than under an evening sun. A bright and deep ring-plain, about 10 miles in diameter, with a distinct central mountain, is connected with the W. wall.

EGEDE.--A lozenge-shaped formation, about 18 miles from corner to corner, bounded by walls scarcely more than 400 feet in height. It is consequently only traceable under very oblique illumination.

THE GREAT ALPINE VALLEY.--A great wedge-shaped depression, cutting through the Alps W. of Plato, from W.N.W. to E.S.E. It is more than 80 miles in length, and varies in breadth from 6 miles on the S. to less than 4 miles on the N., where it approaches the S. border of the Mare Frigoris. For a greater part of its extent it is bounded on the S.W. side by a precipitous linear cliff, which, under a low evening sun, is seen to be fringed by a row of bright little hills. These are traceable up to one of the great mountain masses of the Alps, forming the S.W. side of the great oval-shaped expansion of the valley, whose shape has been appropriately compared to that of a Florence oil-flask, and which Webb terms "a grand amphitheatre." On the opposite or N.E. side, the boundary of the valley is less regular, following a more or less undulating line up to a point opposite, and a little N. of, the great mountain mass, where it abuts on a shallow quasi enclosure with lofty walls, which, projecting westwards, considerably diminish the width of the valley. South of this lies another curved mountain ring, which still farther narrows it. This curtailment in width represents the neck of the flask, and is apparently about 16 or 17 miles in length, and from 3 to 4 miles in breadth, forming a gorge, bordered on the W. by nearly vertical cliffs, towering thousands of feet above the bottom of the valley; and on the E. by many peaked mountains of still greater altitude. At the entrance to the "amphitheatre," the actual distance between the colossal rocks which flank the defile is certainly not much more than 2 miles. From this standpoint the view across the level interior of the elliptical plain would be of extraordinary magnificence. Towards the S., but more than 12 miles distant, the outlook of an observer would be limited by some of the loftiest peaks of the Alps, whose flanks form the boundary of the enclosure, through which, however, by at least three narrow passes he might perchance get a glimpse of the Mare Imbrium beyond. The broadest of these aligns with the axis of the valley. It is hardly more than a mile wide at its commencement on the S. border of the "amphitheatre," but expands rapidly into a trumpet-shaped gorge, flanked on either side by the towering heights of the Alps as it opens out on to the Mare. The bottom, both of the "amphitheatre" and of the long wedge-shaped valley, appears to be perfectly level, and, as regards the central portion of the latter, without visible detail. Under morning illumination I have, however, frequently seen something resembling a ridge partially crossing "the neck," and, near sunset, a tongue of rock jutting out from the E. flank of the constriction, and extending nearly from side to side. At the base of the cliff bordering the valley on the S.W., five or six little circular pits have been noted, some of which appear to have rims. They were seen very perfectly with powers of 350 and 400 on an 8 1/2 inch Calver reflector at 8 h. on January 25, 1885, and have been observed, but less perfectly, on subsequent occasions. The most northerly is about 10 miles from the N.W. end of the formation, and the rest occur at nearly regular intervals between it and "the neck." In the neighbourhood of the valley, on either side, there are several bright craters. Three stand near the N.E. edge, and one of considerable size near the N.W. end on the opposite side. A winding cleft crosses the valley about midway, which, strange to say, is not shown in the maps, though it may be seen in a 4 inch achromatic. It originates apparently at a bright triangular mountain on the plain S.W. of the valley, and, after crossing the latter somewhat obliquely, is lost amid the mountains on the opposite side. That portion of it on the bottom of the valley is easily traceable under a high light as a white line. The region N. of the Alps on the S.W. side of the valley presents many details worthy of examination. Among them, parallel rows of little hills, all extending from N.W. to S.E. There is also a number of still smaller objects of the same type on the E. side. The great Alpine valley, though first described by Schroter, is said to have been discovered on September 22, 1727, by Bianchini, but it is very unlikely that an object which is so prominent when near the terminator was not often remarked before this.

ARCHYTAS.--A bright ring-plain, 21 miles in diameter, on the edge of the Mare Frigoris, due N. of the Alpine Valley, with regular walls rising about 5000 feet above the interior on the N.W., and about 4000 feet on the opposite side. It has a very bright central mountain. Several spurs radiate from the wall on the S., and a wide valley, flanked by lofty heights, forming the S.W. boundary of W.C. Bond, originates on the N side. There is also a crater-rill running towards the N.W. On the Mare, S.W. of Archytas, is a somewhat smaller ring-plain, Archytas A (called by Schmidt, PROTAGORAS), with lofty walls and a central hill.

CHRISTIAN MAYER.--A prominent rhomboidal-shaped ring-plain, 18 miles in diameter, associated on every side, except the N., with a number of irregular inconspicuous enclosures. It has a central peak. Madler discovered two delicate short clefts, both running from N.W. to S.E., one on the W. and the other on the E. of this formation.

W.C. BOND.--A great enclosed plain of rhomboidal shape on the N. of Archytas, the bright ring-plain Timaeus standing near its E. corner, and another conspicuous but much smaller enclosure with a smaller crater W. of it on the floor at the opposite angle. The interior, which is covered with rows of hillocks, is very noteworthy at sunrise.

BARROW.--There are few more striking or beautiful objects at sunrise than this, mainly because of the peculiar shape of its brilliant border and the remarkable shadows of the lofty peaks on its western wall. There is a notable narrow gap in the rampart on the W., which appears to extend to the level of the floor. The walls, especially on the S., are very irregular, and include two large deep craters and some minor depressions. If the formation is observed when its E. wall is on the morning terminator, a fine view is obtained of the remarkable crater-row which winds round the N. side of Goldschmidt. Barrow is about 40 miles in diameter. According to Schmidt, there is one crater in the interior, a little S.E. of the centre.

SCORESBY.--A much fore-shortened deep ring-plain, 36 miles in diameter, between Barrow and the limb. It has a central mountain with two peaks, which are very difficult to detect.

CHALLIS.--A ring-plain adjoining Scoresby on the N.E. It is of about the same size and shape.

MAIN.--A very similar formation, on the N. of the last, much too near the limb to be well observed.

SECOND QUADRANT

EAST LONGITUDE 0 deg. TO 20 deg.

MURCHISON.--A considerable ring-plain about 35 miles across on the E., where it abuts on Pallas. It is a pear-shaped formation, bounded on the N. by a mountainous region, and gradually diminishes in width towards the S.E., on which side it is open to the plain. The walls are of no great altitude, but, except on the N.W., are very bright. At the S. termination of the W. wall there is an exceedingly brilliant crater, Murchison A, five miles in diameter and some 3000 feet deep; adjoining which on the N.W. is an oval depression and a curious forked projection from the border. The only objects visible in the interior are a few low ridges on the E. side, and a number of long spurs running out from the wall on the N. towards the centre of the floor. Murchison A is named CHLADNI by Lohrmann.

PALLAS.--A fine ring-plain, about 32 miles in diameter, forming with Murchison an especially beautiful telescopic object under suitable illumination. Its brilliant border, broken by gaps on the W., where it abuts on Murchison, has a bright crater on the N.E., from which, following the curvature of the wall, and just below its crest, runs a valley in an easterly direction. There is a large bright central mountain on the floor, with a smaller elevation to the S. of it, and a ridge extending from the N. wall to near the centre. On the W., a section of the border is continued in a N. direction far beyond the limits of the formation; and on the S. it is connected with a small incomplete ring; on the E. of which, near the foot of the wall, is a somewhat smaller and much duskier enclosure.

BODE.--A brilliant ring-plain, 9 miles in diameter, situated on the N. side of Pallas. Its walls rise about 5000 feet above the interior, which is considerably depressed, and includes, according to Schmidt and Webb, a mountain or ridge. There are two parallel valleys on the W., which are well worth examination.

SOMMERING.--An incomplete ring-plain, 17 miles in diameter, situated on the lunar equator. It has rather low broken walls and a dark interior.

SCHROTER.--A somewhat larger formation, with a border wanting on the S. Schmidt draws a considerable crater on the S.W. side of the floor. It was in the region north of this object, which abounds in little hills and low ridges, that in the year 1822 Gruithuisen discovered a very remarkable formation consisting of a number of parallel rows of hills branching out (like the veins of a leaf from the midrib) from a central valley at an angle of 45 deg., represented by a depression between two long ridges running from north to south. The regularly arranged hollows between the hills and the longitudinal valley suggested to his fertile imagination that he had at last found a veritable city in the moon--possibly the metropolis of Kepler's Subvolvani, who were supposed to dwell on that hemisphere of our satellite which faces the earth. At any rate, he was firmly convinced that it was the work of intelligent beings, and not due to natural causes. This curious arrangement of ridges and furrows, which, according to Webb, measures about 23 miles both in length and breadth, is, owing to the shallowness of the component hills and valleys, a very difficult object to see in its entirety, as it must be viewed when close to the terminator, and even then the sun's azimuth and good definition do not always combine to afford a satisfactory glimpse of its ramifications. M. Gaudibert has given a drawing of it in the English Mechanic, vol. xviii. p. 638.

GAMBART.--A regular ring-plain, 16 miles in diameter, with a low border and without visible detail within; situated nearly on the lunar equator, about 130 miles S.S.W. of Copernicus, at the N.W. edge of a very hilly region. A prominent pear-shaped mountain, with a small crater upon it, stands a short distance on the S.W., and further in the same direction, a large bright crater with two much smaller craters on the N. of it. The rough hilly district about midway between Copernicus and Gambart is remarkable for its peculiar dusky tone and for certain small dark spots, first seen by Schmidt, and subsequently carefully observed by Dr. Klein. The noteworthy region where these peculiar features are found represents an area of many thousand square miles, and must resemble a veritable Malpais, covered probably with an incalculable number of craters, vents, cones, and pits, filled with volcanic debris. It is among details of this character that the true analogues of some terrestrial volcanoes must be looked for. Under a low angle of illumination the surface presents an extraordinarily rough aspect, well worthy of examination, but the dusky areas and the black spots can only be satisfactorily distinguished under a somewhat high sun. I have, however, seen them fairly well when the W. wall of Reinhold was on the morning terminator.

MARCO POLO.--A small and very irregularly-shaped enclosure (difficult to see satisfactorily) on the S. flank of the Apennines. It is hemmed in on every side by mountains.

ERATOSTHENES.--A noble ring-plain, 38 miles in diameter; a worthy termination of the Apennines. The best view of it is obtained under morning illumination when the interior is about half-filled with shadow. At this phase the many irregular terraces on the inner slope of the E. wall (which rises at one peak 16,000 feet above an interior depressed 8000 feet below the Mare

Imbrium) are seen to the best advantage. The central mountain is made up of two principal peaks, nearly central, from which two bright curved hills extend nearly up to the N.W. wall,--the whole forming a V-shaped arrangement. On the S. there is a narrow break in the wall, and the S.W. section of it seems to overlap and extend some distance beyond the S.E. section. The border on the S.W. is remarkable for the great width of its glacis. Eratosthenes exhibits a marked departure from circularity, especially on the E., where the wall consists of two well-marked linear sections, with an intermediate portion where the crest for 20 miles or more bends inwards or towards the centre. From the S.E. flank of this formation extends towards the W. side of Stadius one of the grandest mountain arms on the moon's visible surface, rising at one place 9000 feet, and in two others 5000 and 3000 feet respectively above the Mare Imbrium. If this magnificent object is observed when the morning terminator falls a little E. of Stadius, it affords a spectacle not easily forgotten. I have often seen it at this phase when its broad mass of shadow extended across the well-known crater-row W. of Copernicus, some of the component craters appearing between the spires of shade representing the loftiest peaks on the mountain arm. There is a prominent little crater on the crest of the arm between two of the peaks, and another on the plain to the west.

STADIUS.--An inconspicuous though a very interesting formation, 43 miles in diameter, W. of Copernicus, with a border scarcely exceeding 200 feet in height. Hence it is not surprising that it was for a long time altogether overlooked by Madler. Except as a known object, it is only traceable under very oblique illumination, and even then some attention is required before its very attenuated wall can be followed all round. It is most prominent on the W., where it apparently consists of a S. extension of the Eratosthenes mountain-arm, and is associated with a number of little craters and pits. This is succeeded on the S.W. by a narrow strip of bright wall, and on the S. by a section made up of a piece of straight wall and a strip curving inwards, forming the S. side. On the E. the border assumes a very ghostly character, and appears to be mainly defined by rows of small depressions and mounds. On the N.E., N., and N.W. it is still lower and narrower; so much so, that it is only for an hour or so after sunrise or before sunset that it can be traced at all. On every side, with the exception of the curved piece on the S., the wall consists of linear sections. The interior contains a great number of little craters and very low longitudinal mounds. Ten craters are shown in Beer and Madler's map. Schmidt only draws fifteen, though in the text accompanying his chart he says that he once counted fifty. In the monograph published in the Journal of the Liverpool Astronomical Society (vol. v. part 8), forty-one are represented. They appear to be rather more numerous on the S. half of the floor than elsewhere. Just beyond the limits of the border on the N., is a bright crater with a much larger obscure depression on the W. of it. The former is surrounded by a multitude of minute craters and crater-cones, which are easily seen under a low sun. Though almost every trace of Stadius disappears under a high light, I have had little difficulty in seeing portions of the border and some of the included details when the morning terminator had advanced as far as the E. wall of Herodotus, and the site was traversed by innumerable light streaks radiating from Copernicus. At this phase the bright crater, just mentioned, on the N. edge of the border was tolerably distinct.

COPERNICUS.--This is without question the grandest object, not only on the second Quadrant, but on the whole visible superficies of the moon. It undoubtedly owes its supremacy partly to its comparative isolation on the surface of a vast plain, where there are no neighbouring formations to vie with it in size and magnificence, but partly also to its favourable position, which is such, that, though not central, is sufficiently removed from the limb to allow all its manifold details to be critically examined without much foreshortening. There are some other formations, Langrenus and Petavius, for example, which, if they were equally well situated, would probably be fully as striking; but, as we see it Copernicus is par excellence the monarch of the lunar ring-mountains. Schmidt remarks that this incomparable object combines nearly all the characteristics of the other ring-plains, and that careful study directed to its unequalled beauties and magnificent form is of much more value than that devoted to a hundred other objects of the same class. It is fully 56 miles in diameter, and, though generally described as nearly circular, exhibits very distinctly under high powers a polygonal outline, approximating very closely to an equilateral hexagon. There are, however, two sections of the crest of the border on the N.E. which are inflected slightly towards the centre, a peculiarity already noticed in the case of Eratosthenes. The walls, tolerably uniform in height, are surmounted by a great number of peaks, one of which on the W., according to Neison, stands 11,000 feet above the floor, and a second on the opposite side is nearly as high. Both the inner and outer slopes of this gigantic rampart are very broad, each being fully 10 miles in width. The outer slope, especially on the E., is a fine object at sunrise, when its rugged surface, traversed by deep gullies, is seen to the best advantage. The terraces and other features on the bright inner declivities on this side may be well observed when the sun's altitude is about 6 deg. Schmidt, whose measures differ from those of Neison, estimates the height of the wall on the E. to be 12,000 feet, and states that the interior slopes vary from 60 deg. to 50 deg. above, to from 10 deg. to 2 deg. at the base. The first inclination of 50 deg., and in some cases of 60 deg., is confined to the loftiest steep crests and to the flanks of the terraces. There are apparently five bright little mountains on the floor, the most easterly being rather the largest, and a great number of minute hillocks on the S.E. quarter. S.W. of the centre is a little crater, and on the same side of the interior a curious hook-shaped ridge, projecting from the foot of the wall, and extending nearly halfway across the floor. The region surrounding Copernicus is one of the most remarkable on the moon, being everywhere traversed by low ridges, enclosing irregular areas, which are covered with innumerable craterlets, hillocks, and other minute features, and by a labyrinth of bright streaks, extending for hundreds of miles on every side, and varying considerably both in width and brilliancy.

The notable crater-row on the W., often utilised by observers for testing the steadiness of the air and the definition of their telescopes, should be examined when it is on the morning terminator, at which time Webb's homely comparison, "a mole-run with holes in it," will be appreciated, and its evident connection with the E. side of Stadius clearly made out. There is another much more delicate row running closely parallel to this object; it lies a little W. of it, and extends farther in a northerly direction.

ARCHIMEDES.--Next to Plato the finest object on the Mare Imbrium. It is about 50 miles in diameter. The average height of its massive border is about 4000 feet above the interior, which is only depressed some 500 or 600 feet below the Mare, the highest peak (about 7000 feet) being on the S.E. The walls are terraced, and include much detail, both within and without. The most noteworthy features in connection with this formation are the crater-cones, craterlets, pits, white spots, and light streaks which figure on the otherwise smooth interior. Mr. T.P. Gray, F.R.A.S., of Bedford, who, with praiseworthy assiduity, has devoted more than ten years to the close scrutiny of these features, Mr. Stanley Williams, and others, have detected four crater-cones on the E. half of the floor, and about fifty minute craters and white spots, also probably volcanic vents, and a very curious and interesting series of light streaks, mostly traversing the formation from E. to W. A little E.

of the centre is a dusky oval area about 6 miles across, and S.W. of this is another, much smaller. Under some conditions of illumination the two principal light markings may be traced over the W. wall, and for some distance on the plain beyond.

On the southern side of Archimedes is a very rugged mountain region, extending for more than 100 miles towards the south: on the W. of this originates a remarkable rill-system, best seen under evening illumination. The two principal clefts follow a nearly parallel course up to the face of the Apennines near Mount Bradley, crossing in their way, almost at right angles, other clefts which run at no great distance from the E. foot of this range and ramify among the outlying hills. Archimedes A is a brilliant little ring-plain on the S.E. of Archimedes. It casts an extraordinary shadow at sunrise, and has a well-marked crater-row on the E. of it, and two long narrow valleys, one of which appears to be a southerly extension of the row.

BEER.--A very bright little crater, with an unnamed formation of about the same size adjoining it on the N., with which is associated a curious winding ridge that appears to traverse a gap in its N. wall.

TIMOCHARIS.--A fine ring-plain, 23 miles in diameter (the centre of a minor ray-system). It stands isolated on the Mare Imbrium (below the level of which it is depressed some 3000 feet), about midway between Archimedes and Lambert. Its walls, rising about 7000 feet above the floor, are conspicuously terraced, and on their W. outer slopes exhibit some remarkable depressions. There is a distinct break on the N., and a bright little crater on the N.W., connected with the foot of the glacis by a prominent ridge. On the bright central mountain, Schmidt, in 1842, detected a crater, which is easily seen under a moderately high light. Timocharis and the neighbourhood, especially the peculiar shape of the terminator on the E. of the formation, is well worth examination at sunrise.

PIAZZI SMYTH.--A conspicuous little ring-plain, 5 or 6 miles in diameter, depressed about 1500 feet below the Mare Imbrium, with a border rising about 2000 feet above it. With the curious arrangement of ridges, of which it is the apparent centre, it is a striking object under a low sun.

KIRCH.--A rather smaller object of the same class on the S.E.

PLATO.--This beautiful walled-plain, 60 miles in diameter, with its bright border and dark steel-grey floor, has, from the time of Hevelius to the present, been one of the most familiar objects to lunar observers. In the rude maps of the seventeenth century it figures as the "Lacus Niger Major," an appellation which not inaptly describes its appearance under a high sun, when the sombre tone of its apparently smooth interior is in striking contrast to that of the isthmus on which the formation stands. It will repay observation under every phase, and though during the last thirty years no portion of the moon has been more diligently scrutinised than the floor; the neighbourhood includes a very great number of objects of every kind, which, not having received so much attention, will afford ample employment to the possessor of a good telescope during many lunations.

The border of Plato, varying in height from 3000 to 4000 feet above the interior, is crowned by several lofty peaks, the highest (7400 feet) standing on the N. side of the curious little triangular formation on the E. wall. Those on the W., three in number, reckoning from N. to S., are respectively about 5000, 6000, and 7000 feet in altitude above the floor. The circumvallation being very much broken and intersected by passes, exhibits many distinct breaches of continuity, especially on the S. There is a remarkable valley on the S.W., which, cutting through the border at a wide angle, suddenly turns towards the S.E., and descends the slope of the glacis in a more attenuated form. Another but shorter valley is traceable at sunrise on the W. On the N.W., the rampart is visibly dislocated, and the gap occupied by an intrusive mountain mass. This dislocation is not confined to the wall, but, under favourable conditions, may be traced across the floor to the broken S.E. border. It is probably a true "fault." On the N.E., the inner slope of the wall is very broad, and affords a fine example of a vast landslip.

The spots and faint light markings on the floor are of a particularly interesting character. During the years 1869 to 1871 they were systematically observed and discussed under the auspices of the Lunar Committee of the British Association. Among the forty or more spots recorded, six were found to be crater-cones. The remainder--or at least most of them--are extremely delicate objects, which vary in visibility in a way that is clearly independent of libration or solar altitude; and, what is also very suggestive, they are always found closely associated with the light markings,--standing either upon the surface of these features or close to their edges. Recent observations of these spots with a 13 inch telescope by Professor W.H. Pickering, under the exceptionally good conditions which prevail at Arequipa, Peru, have revived interest in the subject, for they tend to show that visible changes have taken place in the aspect of the principal crater-cones and of some of the other spots since they were so carefully and zealously scrutinised nearly a quarter of a century ago. The gradual darkening of the floor of Plato as the sun's altitude increases from 20 deg. till after full moon may be regarded as an established fact, though no feasible hypothesis has been advanced to account for it.

On the N.E. of Plato is a large bright crater, A; and, extending in a line from this towards the E., is a number of smaller rings, the whole group being well worth examination. On the N. there is a winding cleft, and some short crossed clefts in the rugged surface just beyond the foot of the wall, which I have seen with a 4 inch achromatic. The region on the W., imperfectly shown in the maps, includes much unrecorded detail. On the Mare Imbrium S. of Plato is a large area enclosed by low ridges, to which Schroter gave the name "Newton." It suggests the idea that it represents the ruin of a once imposing enclosure, of which the conspicuous mountain Pico formed a part.

TIMAEUS.--A very bright ring-plain, 22 miles in diameter, with walls about 4500 feet in height, on the coast-line of the Mare Frigoris, and associated with the E. side of the great enclosed plain W.C. Bond. Schmidt shows a double hill, nearly central, and Neison a crater on the S.W. wall.

BIRMINGHAM.--A large rhomboidal-shaped enclosure, defined by mountain chains and traversed by a number of very remarkable parallel ridges. It is situated nearly due N. of Plato on the N. edge of the Mare Frigoris, and lies on the S.E. side of W.C. Bond, to which it bears a certain resemblance. This region is characterised by the parallelism displayed by many formations, large and small. It is more apparent hereabouts than in any other part of the moon's visible surface. When favourably placed under a low morning sun, Birmingham is a striking telescopic object.

FONTINELLE.--A fine ring-plain, 23 miles in diameter, on the N. margin of the Mare Frigoris, N.N.E. of Plato, with a wall rising on the E., 6000 feet above a bright interior. I find its border indistinct and nebulous, excepting under very oblique light, though three of the little craters upon it are bright and prominent. One stands on the S., another on the N.W., and a third on the E. Schmidt shows only the first of these, and Neison none of them. Fontinelle has a low central mountain which is easily distinguished. Fontinelle A, an isolated mountain on the S., is more than 3000 feet high. On the N. there is a curious mountain group, also of considerable altitude, and on the W. an irregular depression surrounded by a dusky area. North of Fontinelle, extending towards Goldschmidt and the limb, Schroter discovered a very wide irregular valley which he named "J.J. Cassini." It is really nothing more than a great plain bounded by ridges. At 9 h. October 15, 1888, when Philolaus was on the morning terminator, I had a fine view of it, and, as regards its general shape, found that it agreed very closely with Schroter's drawing.

EPIGENES.--A remarkable ring-plain, about 26 miles in diameter, abutting on a mountain ridge running parallel to the E. flank of W.C. Bond. It is a notable object under a low morning sun. There are several elevations on the floor.

GOLDSCHMIDT.--A great abnormally-shaped enclosure with lofty walls between Epigenes and the limb. Neison mentions only two crater-pits within. I have seen the rimmed crater shown by Schmidt on the W. side and three or four other objects of a doubtful kind.

ANAXAGORAS.--A brilliant ring-plain of regular form, 32 miles in diameter, adjoining Goldschmidt on the E. It is a prominent centre of light streaks, some of which traverse the interior of Goldschmidt. On the north a peak rises to the height of 10,000 feet. There is a long ridge on the floor, running from E. to W.

GIOJA.--A ring-plain about 26 miles in diameter, near the north pole.

EAST LONGITUDE 20 deg. TO 40 deg.

REINHOLD.--A prominent ring-plain, 31 miles in diameter, with a lofty border, rising at a peak on the W. to more than 9000 feet above the floor. Its shape on the W. is clearly polygonal, the wall consisting of three rectilineal sections, and on the E. it is made up of two straight sections connected by a curved section. The inner slope includes a remarkably distinct and regular terrace, the E. portion of which is well seen when the interior is about half illuminated by the rising sun. At this phase also the great extent of the glacis on the S.W., and the deep wide gullies traversing it on the E. are observed to the best advantage. The central mountain, though of considerable size, is not prominent. Close to Reinhold on the N.W. stands a noteworthy little formation with a low and partially lineal wall, exhibiting a gap on the north. There is a distinct crater on the S. side of its floor.

GAY-LUSSAC.--A very interesting ring-plain, 15 miles in diameter, situated in the midst of the Carpathian Highlands N. of Copernicus, with a smaller but brighter and deeper formation (Gay-Lussac A) on the S.W. of it, and a conspicuous little crater, not more than 2 or 3 miles in diameter, between the two. The interior of Gay-Lussac is traversed by two coarse clefts, lying nearly in a meridional direction. The more easterly runs from the foot of the S. wall, near the little crater just mentioned, across the floor to the low N. border, which it apparently cuts through, and extends for some distance beyond, terminating in a great oval expansion. The other, which is not shown in the maps, is closely parallel to it, and can be traced up to the N. border, but not farther. Schmidt represents the first as a crater-row, which it probably is, as it varies considerably in width. From the S.E. side of the formation extends a long cleft, terminating at the end of a prominent spur from the S. side of the Carpathians. There are also two remarkable rill-like valleys, commencing on the N. of Gay-Lussac A, which curve round the W. side of Gay-Lussac.

HORTENSIUS.--This brilliant crater, about 10 miles in diameter, is remarkable for its depth, and as being a ray-centre surrounded by a nimbus of light. It has a central mountain, and Schmidt shows a minute crater on the outer slope of the S. wall. The former is a test object.

MILICHIUS.--Is situated on the N.N.E. of Hortensius. It is fully as bright, but rather smaller. Its floor, apparently devoid of detail, is considerably depressed below the surrounding surface.

TOBIAS MAYER.--Like Gay-Lussac, a noteworthy ring-plain associated with the Carpathian Mountains. It is 22 miles in diameter, and has a wall which rises on the W. to a height of nearly 10,000 feet above the floor; on the latter there is a conspicuous central mountain, and on the E. side a crater, and some little hills. Schmidt shows a smaller crater on the W. side, which I have not seen. Adjoining the formation on the W. is a ring- plain of about one-fourth its area, which is a bright object. Tobias Mayer and the neighbouring Carpathians form an especially beautiful telescopic picture at sunrise.

KUNOWSKY.--An inconspicuous ring-plain, about 11 miles in diameter, standing in a barren region in the Mare Procellarum, W.S.W. of Encke. The central mountain is tolerably distinct.

ENCKE.--A regular ring-plain, 20 miles in diameter, with a comparatively low border, nowhere rising more than 1800 feet above the interior, which is depressed some 1000 feet below the surrounding Oceanus Procellarum. A lofty ridge traverses the floor from S. to N., bifurcating before it reaches the N. wall. There is a bright crater on the W. wall, and a depression on the opposite wall, neither of which, strange to say, is shown on the maps. Encke is encircled by ridges, which, when it is on the morning terminator, combine to make it resemble a large crater surrounded by a vast mountain ring.

KEPLER.--One of the most brilliant objects in the second quadrant,--a ring-plain about 22 miles in diameter, with a lofty border; a peak on the E. attaining an altitude of 10,000 feet above the surface. The wall is much terraced, especially the outer slope on the W., where a narrow valley is easily traceable. Though omitted from the maps, there is a prominent circular depression on the W. border, which forms a distinct notch thereon at sunrise. On the N., the wall exhibits a conspicuous gap. There is a central hill on the floor. Immediately E. of Kepler is a bright plateau, bounded on the N. by a very straight border, with two small craters on its edge. Both these objects are incomplete on the N., as if they had been deformed by a "fault," which has apparently affected the N. end of Kepler also. Kepler is the centre of one of the most extended systems of bright streaks on the moon's visible surface.

BESSARION.--A bright little ring-plain, about 6 miles in diameter, in the Oceanus Procellarum N. of Kepler. There is a smaller and still brighter companion on the N. (Bessarion E), standing on a light area. Bessarion has a minute central hill, difficult to detect.

PYTHEAS.--A small rhomboidal-shaped ring-plain, 12 miles in diameter, standing in an isolated position on the Mare Imbrium between Lambert and Gay-Lussac. Its bright walls, rising about 2500 feet above the Mare, are much terraced within, especially on the E. There is a bright little crater on the N. outer slope, with a short serpentine ridge running up to it from the region S. of Lambert, and another winding ridge extending from the S. wall to the E. of two conspicuous craters, standing about midway between Pytheas and Gay-Lussac. The former bears a great resemblance to the ridge N. of Madler, and, like this, appears to traverse the N. border. The interior of Pytheas, which is depressed more than 2000 feet below the Mare, includes a brilliant central peak.

LAMBERT.--A ring-plain, 17 miles in diameter, presenting many noteworthy features. The crest of its border stands about 2000 feet above the Mare Imbrium, and more than double this height above the interior. The wall is prominently terraced both within and without; the outer slope on the W. exhibiting at sunrise a nearly continuous valley running round it. When near the morning terminator, the region on the N. is seen to be traversed by some very remarkable ridges and markings; one cutting across the N. wall appears to represent a "fault." On the S. is a large polygonal enclosure formed by low ridges. On the W., towards Timocharis, is a brilliant mountain 3000 feet high, a beautiful little object under a low sun.

LEVERRIER.--The more westerly of a pair of little ring-plains on the N. side of the Mare Imbrium, and S.W. of the Laplace promontory. It is about 10 miles in diameter, with walls rising some 1500 feet above the Mare, and more than 6000 feet above the interior, which seems to be without a central mountain or other features. Schmidt shows the crater on the N. rim and another on the S.E. slope, both of which are omitted by Neison, though they are easy objects when Helicon is on the morning terminator. About 20 miles on the S.E. there is a very bright little crater on a faint light area.

HELICON.--The companion ring-plain on the E. It is 13 miles in diameter, and is very similar, though not quite so deep. There is a crater on the S.E. wall, and, according to Neison, another on the outer slope of the N. border. Webb records a central crater. If Helicon is observed when on the morning terminator, it will be seen to be traversed by a curved ridge which cuts through the walls, and runs up to a bright crater S.E. of Leverrier. It appears to be a "fault," whose "downthrow," though slight, is clearly indicated by an area of lower ground on the E. There is a great number of small craters in the neighbourhood of this formation.

EULER.--The most easterly of the row of great ring-plains, which, beginning on the W. with Autolycus, and followed by Archimedes, Timocharis, and Lambert, extends almost in a great circle from the N.W. to the S.E. side of the Mare Imbrium. It is about 19 miles in diameter, and is surrounded by terraced walls, which, though of no great height above the Mare, rise 6000 feet above the floor. There is a distinct little gap in the S. wall, easily glimpsed when it is close to the morning terminator, which probably represents a small crater. Euler has a bright central mountain, and is a centre of white silvery streaks.

BRAYLEY.--A very conspicuous little ring-plain E.S.E. of Euler, with two smaller but equally brilliant objects of the same class situated respectively E. and W. of it.

DIOPHANTUS.--Forms with Delisle, its companion on the N., a noteworthy object. It is about 13 miles in diameter, with a wall, which has a distinct break in its continuity on the N., rising about 2500 feet above the Mare. A rill-valley runs from the E. side of the ring towards the W. face of a triangular-shaped mountain on the E. of a line joining the formation with Delisle. North are three bright little craters in a line, the middle one being much the largest. From the most easterly of these objects a light streak may be traced under a high sun, extending for many miles to another small crater on the N.W. of Diophantus, and expanding at a point due N. of the formation into a spindle-shaped marking. At sunrise, the W. portion of the streak has all the appearance of a cleft, with a branch about midway running to the S. side of Delisle. Under the same phase a broad band of shadow extends from the N.E. wall to the triangular mountain just mentioned, representing a very sudden drop in the surface--resembling on a small scale the well-known "railroad" E. of Thebit. Diophantus has no central mountain.

DELISLE.--A larger and more irregularly-shaped object than the last, 16 miles in diameter, with loftier and more massive walls, and an extensive but ill-defined central hill. There is an evident break in the northern border. A triangular mountain on the S.E. and a winding ridge running up to the N. wall are prominent features at sunrise, as are also the brilliant summits of a group of hills some distance to the E.N.E.

CARLINI.--A small but prominent and deep little crater about 5 miles in diameter on the Mare Imbrium about midway between Lambert and the Sinus Iridum. There are many faint light streaks in the vicinity, one of which extends from Carlini to Bianchini, on the edge of the Sinus,--a distance of 300 miles. Schmidt shows a central peak.

CAROLINE HERSCHEL.--A bright and very deep ring-plain about 8 miles in diameter on the Mare Imbrium, some distance E.N.E. of the last. On the S.E. lies a larger crater, Delisle B, which has a small but obvious crater on its N. rim, and casts a very prominent shadow at sunrise. Caroline Herschel stands on a long curved ridge running N.E. from Lambert towards the region E. of Helicon, and, according to Schmidt, has a central peak. On the E. is a bright mountain with two peaks; some distance N. of which is a large ill-defined white spot, with another spot of a similar kind on the W. of it, nearly due N. of Caroline Herschel.

GRUITHUISEN.--This ring-plain, 10 miles in diameter, is situated on the Mare Imbrium on the N.E. of Delisle. It is associated with a number of ridges trending towards the region N. of Aristarchus and Herodotus.

THE LAPLACE PROMONTORY.--A magnificent headland marking the extreme western extremity of the finest bay on the moon's visible surface, the Sinus Iridum; above which it towers to a height of 9000 feet or more, projecting considerably in front of the line of massive cliffs which define the border of the Sinus, and presenting a long straight face to the S.E. Near its summit are two large but shallow depressions, the more easterly having a very bright interior. At a lower level, almost directly below the last, is a third depression. All three are easy objects under a low sun. The best view of the promontory and its surroundings is obtained when the E. side of the bay is on the morning terminator. Its prominent shadow is traceable for many days after sunrise.

THE HERACLIDES PROMONTORY.--The less lofty but still very imposing headland at the E. end of the Sinus Iridum, rising more than 4000 feet above it. It consists of a number of distinct mountains, forming a triangular-shaped group running out to a point at the S.W. extremity of the bay, and projecting considerably beyond the shore-line. There is a considerable crater on the E. side of the headland, not visible till a late stage of sunrise. It is among the mountains composing this promontory that some ingenuity and imagination have been expended in endeavouring to trace the lineaments of a female face, termed the "Moon-maiden."

BIANCHINI.--A fine ring-plain, about 18 miles in diameter, on the N.E. side of the Sinus Iridum, surrounded by the lofty mountains defining the border of the bay. Its walls, which are prominently terraced within, rise about 7000 feet on the E., and about 8000 feet on the W. above the floor, which includes a prominent ridge and a conspicuous central mountain. There is a distinct crater on the S. wall, not shown in the maps. Between this side of the formation and the bay is a number of hills running parallel to the shore-line: these, with the intervening valleys, will repay examination at sunrise.

MAUPERTUIS.--A great mountain enclosure of irregular shape, about 20 miles in diameter, in the midst of the Sinus Iridum highlands, N. of Laplace. The walls are much broken by passes, and the interior includes many hills and ridges.

CONDAMINE.--A rhomboidal-shaped ring-plain, about 23 miles in diameter, N. of Maupertuis, with lofty walls, especially on the E., where they rise some 4000 feet above the interior. There are three large depressions on the outer N.W. slope, and at least three minute craters on the crest of the wall just above. Though neither Neison nor Schmidt draw any detail thereon, there is a prominent ridge on the N. side of the floor, and a low circular hill on the S. On the S.E. four long ridges or spurs radiate from the wall, and on the N.E. are three remarkable square-shaped enclosures. On the edge of the Mare Frigoris, N.W. of Condamine, are many little craters with bright rims and a distinct short cleft, running parallel to the coast-line. The winding valleys in the region bordering the Sinus Iridum, and other curious details, render this portion of the moon's surface almost unique.

BOUGUER.--A bright regular little ring-plain, about 8 miles in diameter, N. of Bianchini.

J.F.W. HERSCHEL.--A vast enclosed plain, about 90 miles across, bounded on the W. by a mountain range, which here defines the E. side of the Mare Frigoris, on the S. by massive mountains, and on the other sides by a lofty but much broken wall, intersected by many passes. Within is a large ring-plain, nearly central, and a large number of little craters and crater-pits. The floor is traversed longitudinally by many low ridges, lying very close together, which at sunrise resemble fine grooves or scratches of irregular width and depth.

HORREBOW.--A ring-plain of remarkable shape, resembling the analemma figure, standing at the S. end of the mountain range bounding J.F.W. Herschel on the W. Schmidt shows a crater on the W. wall, near the constriction on this side, and a second at the foot of the slope of the E. wall.

PHILOLAUS.--A ring-plain 46 miles in diameter, on the N.E. of Fontinelle. Its bright walls rise on the W. to a height of nearly 12,000 feet above the floor (on which there is a conspicuous central mountain), and exhibit many prominent terraces. Philolaus is partially encircled, at no great distance, by a curved ridge, on which will be found a number of small craters.

ANAXIMINES.--A much foreshortened ring-plain, about 66 miles in diameter, on the E. of Philolaus. One peak on the E. is nearly 8000 feet in height. Schmidt shows four craters on the W. side of the floor, and a fifth on the S.E. side. There is a bright streak in the interior, which extends southwards for some distance across the Mare Frigoris.

EAST LONGITUDE 40 deg. TO 60 deg.

REINER.--A regular ring-plain, 21 miles in diameter, in the Mare Procellarum, S.S.E. of Marius, with a very lofty border terraced without and within, and a minute but conspicuous mountain standing at the N. end of a ridge which traverses the uniformly dark floor in a meridional direction. A long ridge extends some way towards the S. from the foot of the S. wall, and at some distance in the same direction lie six ill- defined white spots of doubtful nature. On the E.N.E. there is a large white marking, resembling a "Jew's harp" in shape, and farther on, towards the E., a number of very remarkable ridges. On the W. will be found many bright little craterlets. A ray from Kepler extends almost up to the W. wall of Reiner.

MARIUS.--A very noteworthy ring-plain, 27 miles in diameter, in the Oceanus Procellarum, E.N.E. of Kepler, with a bright border rising about 4000 feet above the interior, which is of an uneven tone. The rampart exhibits some breaks, especially on the S. The outer slope on the W. is traversed by a fine deep valley, distinctly marked when the opposite side is on the morning terminator. It originates on the S.W. at a prominent crater situated a little below the crest of the wall, and, following its curvature, runs out on to the plain near a large mountain just beyond the foot of the N. border. In addition to the crater just mentioned, there are two smaller ones below the summit of the S. wall, and a small circular depression on the S.E. wall. Mr. W.H. Maw, F.R.A.S., has seen, with a 6 inch Cooke refractor, a bright marking at the N. extremity of the ring, which, when examined with a Dawes' eyepiece, resembled an imperfect crater. The floor includes at least four objects--(1) A crater on the N.W., standing on a circular light area; (2) a white spot a little S. of the centre; (3) a smaller white spot S.E. of this; (4) another, near the inner foot of the S.W. wall. Marius is an imposing object under oblique illumination, mainly because of the number of ridges by which it is surrounded. I have frequently remarked at sunrise that the surface on the W., and especially the outer slope of the rampart, is of a decided brown or sepia tint, similar to that which has already been noticed with respect to Geminus and its vicinity, viewed under like conditions. Schmidt in 1862 discovered a long serpentine cleft some distance N. of Marius, which has not been seen since.

ARISTARCHUS.--The brightest object on the moon, forming with Herodotus (a companion ring-plain on the E.), and its remarkable surroundings, one of the most striking objects which the telescope has revealed on the visible surface, and one requiring much patient observation before its manifold details can be fully noted and duly appreciated. Its border rises 2000 feet above the outer surface on the W., but towers to more than double this height above the glistening floor. No lunar object of its moderate dimensions (it is only about 29 miles in diameter) has such conspicuously terraced walls, or a greater number of spurs and buttresses; which are especially prominent on the S. A valley runs round the outer slope of the W. wall, very similar to that found in a similar position round Marius. There is also a distinct valley on the brilliant inner slope of the E. wall, below its crest. It originates at a bright little crater, and is traceable round the greater portion of the declivity. Under a moderately high sun, an oval area, nearly as large and fully as brilliant as the central mountain, is seen on this inner slope. It is bordered on either side by bands of a duskier hue, which probably represent shallow transverse valleys. From its dazzling brilliancy it is very difficult to observe the interior satisfactorily. In addition, however, to the central mountain, there is a crater on the N.W. side of the floor. On the S. side of Aristarchus is a large dusky ring some 10 miles in diameter, connected by ridges with the spurs from the wall, and on the S.E., close to the foot of the slope, is another smaller ring of a like kind.

HERODOTUS.--This far less brilliant but equally interesting object is about 23 miles in diameter, and is not so regular in shape as Aristarchus. Its W. wall rises at one point more than 4000 feet above the very dusky floor. Except on the S.W. and N.E., the border is devoid of detail. On the S.W. three little notches may be detected on its summit, which probably represent small craters, while on the opposite side, on the inner slope, a little below the crest, is a large crater, easily seen. Both the E. and W. sections of the wall are prolonged towards the S. far beyond the limits of the formation. These rocky masses, with an intermediate wall, are very conspicuous under oblique illumination, that on the S.W. being especially brilliant. On the N. there is a gap through which the well-known serpentine cleft passes on to the floor. Between the N.W. side of Herodotus and Aristarchus is a large plateau, seen to the best advantage when the morning terminator lies a little distance E. of the former. It is traversed by a T-shaped cleft which communicates with the great serpentine cleft and extends towards the S. end of Aristarchus, till it meets a second cleft (forming the upper part of the T) running from the S.E. side of this formation along the W. side of Herodotus. The great serpentine cleft, discovered by Schroter, October 7, 1787, is in many respects the most interesting object of its class. It commences at the N. end of a short wide valley, traversing mountains some distance N.E. of Herodotus, as a comparatively delicate cleft. After following a somewhat irregular course towards the N.W. for about 50 miles, and becoming gradually wider and deeper, it makes a sudden turn and runs for about 10 miles in a S.W. direction. It then changes its course as abruptly to the N.W. again for 3 or 4 miles, once more turns to the S.W., and, as a much coarser chasm, maintains this direction for about 20 miles, till it reaches the S.E. edge of a great mountain plateau N. of Aristarchus, when it swerves slightly towards the S., becoming wider and wider, up to a place a few miles N. of Herodotus, where it expands into a broad valley; and then, somewhat suddenly contracting in width, and becoming less coarse, enters the ring-plain through a gap in the N. wall, as before mentioned. I always find that portion of the valley in the neighbourhood of Herodotus more or less indistinct, though it is broad and deep. This part of it, unless it is observed at a late stage of sunrise, is obscured by the shadow of the mountains on the border of the plateau. Gruithuisen suspected a cleft crossing the region embraced by the serpentine valley, forming a connection between its coarse southern extremity and the long straight section. This has been often searched for, but never found. It may exist, nevertheless, for in many instances Gruithuisen's discoveries, though for a long time discredited, have been confirmed. The mountain plateau N. of Aristarchus deserves careful scrutiny, as it abounds in detail and includes many short clefts.

HARBINGER, MOUNTAINS.--A remarkable group of moderate height, mostly extending from the N.W. towards Aristarchus. They include a large incomplete walled-plain about 30 miles in diameter, defined on the W. by a lofty border, forming part of a mountain chain, and open to the south. This curious formation has many depressions in connection with its N.W. edge. On the N. of it there is a crater-row and a very peculiar zig-zag cleft. The region should be observed when the E. longitude of the morning terminator is about 45 deg.

SCHIAPARELLI.--A conspicuous formation, about 16 miles in diameter, between Herodotus and the N.E. limb, with a border rising nearly 2000 feet above the Mare, and about 1000 more above the floor, on which Schmidt shows a central hill.

WOLLASTON.--A small bright crater on the Mare N. of the Harbinger Mountains, surpassed in interest by a remarkable formation a few miles S. of it, Wollaston B, an object of about the same size, but which is associated with a much larger enclosure, resembling a walled-plain, lying on the N. side of it. This formation has a lofty border on the W., surmounted by two small craters. The wall is lower on the E. and exhibits a gap. There is a central hill, only visible under a low sun. About midway between Wollaston and this enclosure stands a small isolated triangular mountain. From a hill on the E. runs a rill valley to the more westerly of a pair of craters, connected by a ridge, on the S.E. of Wollaston B.

MAIRAN.--A bright ring-plain of irregular shape, 25 miles in diameter, on the E. of the Heraclides promontory. The border, especially on the E., varies considerably in altitude, as is evident from its shadow at sunrise; at one peak on the W. it is said to attain a height of more than 15,000 feet above the interior. There is a very minute crater on the crest of the S. wall, down the inner slope of which runs a rill-like valley. About halfway down the inner face of the E. wall are two other small craters, connected together by a winding valley. These features may be seen under morning illumination, when about one-fourth of the floor is in sunlight. Schroter is the only selenographer who gives Mairan a central mountain. In this he is right. I have seen without difficulty on several occasions a low hill near the centre. The formation is surrounded by a number of conspicuous craters and crater-pits. On the N. there is a short rill-like valley, and another, much coarser, on the S.

SHARP.--A ring-plain somewhat smaller than the last, on the E. of the Sinus Iridum, from the coast-line of which it is separated by lofty mountains. There is a distinct crater at the foot of its N.E. wall, and a bright central mountain on the floor. On the N. is a prominent enclosure, nearly as large as Sharp itself; and on the N.E. a brilliant little ring- plain, A, about 8 miles in diameter, connected with Sharp, as Madler shows, by a wide valley.

LOUVILLE.--A triangular-shaped formation on the E. of a line joining Mairan and Sharp. It is hemmed in by mountains, one of which towers 5000 feet above its dusky floor.

FOUCAULT.--A bright deep ring-plain, about 10 miles in diameter, lying E. of the mountains fringing the Sinus Iridum, between Bianchini and Harpalus. A very lofty peak rises near its N. border, and, according to Neison, it has a distinct central mountain, though neither Madler or Schmidt show any detail within.

HARPALUS.--A conspicuous ring-plain, about 14 miles in diameter, on the N.E. of the last, with a floor sinking 13,000 feet below the surrounding surface. As the cubic contents of the border and glacis are quite inadequate to account for it, we may ask, what has become of the material which presumably once filled this vast depression? Harpalus has a bright central mountain.

SOUTH.--On the W. and S., the boundaries of this extensive enclosure are merely indicated by ridges, which nowhere attain the dignity of a wall. On the N., the edge of a tableland intersected by a number of valleys define its limits, and on the E. a border forming also the W. side of Babbage. The interior is traversed by a number of longitudinal hills, and includes two bright little heights, drawn by Schmidt as craters.

BABBAGE.--A still larger enclosed area, adjoining South on the E., and containing a considerable ring-plain near its W. border. It is a fine telescopic object at sunrise, the interior being crossed by a number of transverse markings representing ridges. These are very similar in character (but much coarser) to those found on the floor of J.F.W. Herschel. The curious detail on the E. wall is also worth examination at this phase.

ROBINSON.--A bright and very deep little ring-plain, about 12 miles in diameter, on a plateau N. of South. Schmidt shows a crater on the W. border, and two others at the foot of the N. and E. borders respectively.

ANAXIMANDER.--A fine but much foreshortened ring-plain, 39 miles in diameter, abutting on the E. side of J.F.W. Herschel. It has a large crater on its W. border, which is common to the two formations, and a very prominent crater, both on the S. and N. The barrier on the S.W. rises to a height of nearly 10,000 feet. Schmidt shows a crater and other details on the floor.

EAST LONGITUDE 60 deg. TO 90 deg.

LOHRMANN.--This ring-plain, with Hevel and Cavalerius on the N. of it, is a member of a linear group, which, but for its propinquity to the limb, would be one of the most imposing on the moon's surface. Lohrmann, about 28 miles in diameter, is surrounded by a bright wall, which, to all appearance, is devoid of detail. Two prominent ridges, with a fine intervening valley, connect it with the N. end of Grimaldi. It has a large but not conspicuous central mountain. On the rugged surface, between the ring-plain and the E. edge of the Oceanus Procellarum, lies a very interesting group of crossed clefts, some of which run from N.E. to S.W., and others from N.W. to S.E. Three of the latter proceed from different points in a coarse valley extending from W. to E., and cross the ridges just mentioned. They follow a parallel course, and terminate on the S. side of a crater-row, consisting of three bright craters ranging in a line parallel to the coarse valley. On the N. side of these objects, and tangential to them, is another cleft, which traverses the W. and E. walls of Lohrmann, and, crossing the region between it and Riccioli, terminates apparently at the W. wall of the latter formation. No map shows this cleft, though it is obvious enough; and, when the E. wall of Hevel is on the morning terminator, the notches made by it in the border of Lohrmann are easily detected. Capt. Noble, F.R.A.S., aptly compares two of the crossed clefts to a pair of scissors, the craters at which they terminate representing the oval handles. On the grey surface of the Mare W. of Lohrmann is the bright crater Lohrmann A, from which, running N., proceeds a rill-like valley ending at a large white spot, which has a glistening lustre under a high light.

HEVEL.--A great walled-plain, 71 miles in diameter, adjoining Lohrmann on the N., with a broad western rampart, rising at one peak to a height above the interior of nearly 6000 feet, and presenting a steep bright face to the Oceanus Procellarum. There are three prominent craters near its crest, and one or two breaks in its continuity. It is not so lofty and is more broken on the E., where three conspicuous craters stand on its inner slope. The floor is slightly convex, and includes a triangular central mountain, on which there is a small crater. The S. half of the interior is crossed by four clefts: (I) running from a little crater N. of the central mountain, on the W. side of it, to a hill at the foot of the S.W. wall; (2) originating near the most southerly of the three craters on the inner slope of the E. wall, and crossing 1, terminates at the foot of the W. wall; (3) has the same origin as 2, crosses 1, and, passing over a craterlet W. of the central mountain, also runs up to the W. wall at a point considerably N. of that where 2 joins the latter; (4) runs from the craterlet just mentioned to the W. end of 2.

CAVALERIUS.--The most northerly member of the linear chain, a ring-plain, 41 miles in diameter, with terraced walls rising about 10,000 feet above the floor. Within there is a long central mountain with three peaks. Under a high light the region on the W. is seen to be crossed by broad light streaks.

OLBERS.--A large ring-plain, 41 miles in diameter, near the limb, N.E. of Cavalerius. Though a very distinct formation, it is difficult to see its details except under favourable conditions of libration. It has a large crater on its W. wall, a smaller one on the E., and a third on the N. The floor includes a central mountain, and, according to Schmidt, four craters. He also shows a crater-rill on the W. wall, N. of the large crater thereon. Olbers is the origin of a fine system of light rays.

GALILEO.--A ring-plain, about 9 miles in diameter, N.E. of Reiner, associated with ridges, some of which extend to the "Jew's Harp" marking referred to under this formation.

CARDANUS.--A fine regular ring-plain, about 32 miles in diameter, near the limb N. of Olbers. Its bright walls, rising about 4000 feet above the light grey floor, are clearly terraced, and exhibit, especially on the S.E., several spurs and buttresses. There is a fine valley on the outer W. slope, a large bright crater on the Mare just beyond its foot, and a conspicuous mountain in the same position farther north. I have not succeeded in seeing the faint central hill nor the crater N. of it shown by Schmidt, but there is a brilliant white circular spot on the floor at the inner foot of the N.E. wall which he does not show.

KRAFFT.--A very similar object on the N., of about the same dimensions; with a central peak, and a large crater on the dark floor abutting on the S.W. wall, and another of about half the size on the outer side of the W. border. From this crater a very remarkable cleft runs to the N. wall of Cardanus: it is bordered on either side by a bright bank, and cuts through the N.W. border of the latter formation. It is best seen when the E. wall of Cardanus is on the morning terminator.

VASCO DE GAMA.--A bright enclosure, 51 miles in diameter, with a small central mountain. It is associated on the N. with a number of enclosed areas of a similar class, all too near the limb to be well seen.

SELEUCUS.--A considerable ring-plain, 32 miles in diameter, with lofty terraced walls, rising 10,000 feet above a dark floor which includes an inconspicuous central hill. This formation stands on a ridge extending from near Briggs to the W. side of Cardanus.

OTTO STRUVE.--An enormous enclosure, bounded on the E. by the Hercynian Mountains, and on the W. by a mountain chain of considerable altitude, surmounted by three or more bright little rings. On the W. side of the uneven-toned interior, which, according to Madler, includes an area of more than 26,000 square miles, stand four craters, several little hills, and light spots. On the W. is the much more regular and almost as large formation, Otto Struve A, the W. border of Otto Struve forming its E. wall. This enclosure is bounded elsewhere by a very low, broken, and attenuated barrier. At sunrise the E. and W. walls, with the mountain mass at the N. end, which they join, resemble a pair of partially-opened calipers. There is one conspicuous little crater on the W. side of the floor; and, at or near full moon, four or five white spots, nearly central, are prominently visible.

BRIGGS.--This bright regular ring-plain, 33 miles in diameter, is situated a short distance N. of Otto Struve A. A long ridge traverses the interior from N. to S. On the E. is another large enclosure, communicating with Otto Struve on the S., and really forming a N. extension of this formation. It has a large and very deep crater, 12 miles in diameter, on its W. border.

LICHTENBERG.--A conspicuous little ring-plain, about 12 miles in diameter, in an isolated position on the Mare, some distance N. of Briggs. It was here that Madler records having occasionally noticed a pale reddish tint, which, though often searched for, has not been subsequently seen.

ULUGH BEIGH.--A good-sized ring-plain, E. of the last, with a bright border and central mountain. Too near the limb for observation.

LAVOISIER.--A small bright walled-plain N. of Ulugh Beigh. It has a somewhat dark interior. West of it is Lavoisier A, a ring-plain about 14 miles in diameter. Both are too near the limb for useful observation.

GERARD.--A large enclosure close to the limb, still farther N., containing a long ridge and a crater.

HARDING.--A small ring-plain W. of Gerard, remarkable for the peculiar form of its shadow at sunrise, and for the ridges in its vicinity.

REPSOLD.--The largest of a group of walled enclosures, close to the limb, on the E. side of the Sinus Roris.

XENOPHANES.--But for its position, this deep walled-plain, 185 miles in diameter, would be a fine telescopic object, with its lofty walls, large central mountain, and other details.

OENOPIDES.--A large and tolerably regular walled-plain, 43 miles in diameter, on the W. of the last. The depressions on the W. wall are worth examination at sunrise. There is apparently no detail whatever on the floor.

CLEOSTRATUS.--A small ring-plain, N. of Xenophanes, surrounded by a number of similar objects, all too near the limb for observation.

PYTHAGORAS.--A noble walled-plain, 95 miles in diameter, which no one who observes it fails to lament is not nearer the centre of the disc, as it would then undoubtedly rank among the most imposing objects of its class. Even under all the disadvantages of position, it is by far the most striking formation in the neighbourhood. Its rampart rises, at one point on the N., to a height of nearly 17,000 feet above the floor, on which stands a magnificent central mountain, familiar to most observers.

THIRD QUADRANT

EAST LONGITUDE 0 deg. TO 20 deg.

MOSTING.--A very deep ring-plain, 15 miles in diameter, near the moon's equator, and about 6 deg. E. of the first meridian. There is a crater on the N. side of its otherwise unbroken bright border, an inconspicuous central mountain, and, according to Neison, a dark spot on the S. side of the floor. At some distance on the S.S.W., stands the bright crater, Mosting A, one of the most brilliant objects on the moon's visible surface.

REAUMUR.--A large pentagonal enclosure, about 30 miles in diameter, with a greatly broken border, exhibiting many wide gaps, situated on the E. side of the Sinus Medii, N.W. of Herschel. The walls are loftiest on the S. and S.W., where several small craters are associated with them. A ridge connects the formation with the great deep crater Reaumur A, and a second large enclosure lying on the W. side of the well-known valley W. of Herschel. At the end of a spur on the S. side of the great crater originates a cleft, which I have often traced to the N.W. wall of Ptolemaeus, and across the N. side of the floor of this formation to a crater on the N.E. quarter of it, Ptolemaeus d. There is a short cleft on the W. side of the floor of Reaumur, running from N. to S.

HERSCHEL.--A typical ring-plain, situated just outside the N. border of Ptolemaeus, with a lofty wall rising nearly 10,000 feet above a somewhat dusky floor, which includes a prominent central mountain. Its bright border is clearly terraced both within and without, the terraces on the inner slope of the W. wall being beautifully distinct even under a high light, and on the outer slope are some curious irregular depressions. On the S.S.E. is a large oblong deep crater, close to the rocky margin of Ptolemaeus, and a little beyond the foot of the wall on the N.W. is a smaller and more regular rimmed depression, b, standing near the E. border of the great valley, more than 80 miles long, and in places fully 10 miles wide, which runs from S.S.W. to N.N.E. on the W. side of Herschel, and bears a close resemblance to the well-known Ukert Valley. Herschel d is a large but shallow ring-plain on the E. of Herschel, with a brilliant but smaller crater on the W. of it.

North of Herschel, on a plateau concentric with its outline, stands the large polygonal ring-plain Herschel a, a formation of a very interesting character, with a low broken wall, exhibiting many gaps, and including some craters of a minute class. The largest of these stands on the S.W. wall. Mr. W.H. Maw has detected some of these objects on the N. side, both in connection with the border and beyond it.

FLAMMARION.--A large incomplete walled-plain N.E. of Herschel, open towards the N., with a border rising about 3000 feet above the floor. The brilliant crater, Mosting A, stands just outside the wall on the E.

PTOLEMAEUS.--Taking its very favourable position into account, this is undoubtedly the most perfect example of a walled-plain on the moon's visible superficies. It is the largest and most northerly component of the fine linear chain of great enclosures, which extend southwards, in a nearly unbroken line, to Walter. It exhibits a very marked departure from circularity, the outline of the border approximating in form to a hexagon with nearly straight sides. It includes an area of about 9000 square miles, the greatest distance from side to side being about 115 miles. It is, in fact, about equal in size to the counties of York, Lancashire, and Westmorland combined; and were it possible for one to stand near the centre of its vast floor, he might easily suppose that he was stationed on a boundless plain; for, except towards the west, not a peak, or other indication of the existence of the massive rampart would be discernible; and even in this direction he would only see the upper portion of a great mountain on the wall.

The border is much broken by gaps and intersected by passes, especially E. and S., where there are several valleys connecting the interior with that of Alphonsus. The loftiest portion of the wall, which includes many crateriform depressions, is on the W., where one peak rises to nearly 9000 feet. Another on the N.E. is about 6000 feet above the interior. On the N.W. is a remarkable crater-row, called, from its discoverer, "Webb's furrow," running from a point a little N. of a depression on the border to a larger crateriform depression on the S. of Hipparchus K. Birt terms it "a very fugitive and delicate lunar feature." As regards the vast superficies enclosed by this irregular border, it is chiefly remarkable for the number of large saucer-shaped hollows which are revealed on its surface under a low sun. They are mostly found on the eastern quarter of the floor. Some of them appear to have very slight rims, and in two or three instances small craters may be detected within them. Owing to their shallowness, they are very evanescent, and can only be glimpsed for an hour or so about sunrise or sunset. The large bright crater A, about 4 miles in diameter on the N.W. side of the interior, is by far the most conspicuous object upon it. Adjoining it on the N. is a large ring with a low border, and N. of this again is another, extending to the wall. Mr. Maw and Mr. Mee have seen minute craters on the borders of these obscure formations. In addition to the objects just specified, there is a fairly conspicuous crater, d, on the N.E. quarter of the floor, and a very large number of others distributed on its surface, which is also traversed by a network of light streaks, that have recently been carefully recorded by Mr. A.S. Williams. A cleft, from near Reaumur A, traverses the N. side of the floor, and runs up to Ptolemaeus d.

ALPHONSUS.--A large walled-plain, 83 miles in diameter, with a massive irregular border abutting on the S.S.E. side of Ptolemaeus, and rising at one place on the N.W. to a height of 7000 feet above the interior. The floor presents many features of interest. It includes a bright central peak, forming part of a longitudinal ridge, on either side of which runs a winding cleft, originating at a crater-row on the N. side of the interior. There is a third cleft on the N.W. side, and a fourth near the foot of the E. wall. There are also three peculiar dark areas within the circumvallation; two, some distance apart, abutting on the W. wall, and a third, triangular in shape, at the foot of the E. wall. The last- mentioned cleft traverses this patch. These dusky spots are easily recognised in good photographs of the moon.

ALPETRAGIUS.--A fine object, 27 miles in diameter, closely connected with the S.E. side of Alphonsus. It has peaks on the W. towering 12,000 feet above the floor, on which there is an immense central mountain, which in extent, complexity, and altitude surpasses many terrestrial mountain systems--as, for example, the Snowdonian group. The massive barrier between Alpetragius and Alphonsus deserves careful scrutiny, and should be examined under a moderately low morning sun. On the E., towards Lassell, stands a brilliant light-surrounded crater.

ARZACHEL.--Another magnificent object, associated on the N. with Alphonsus, about 66 miles in diameter, and encircled by a massive complex rampart, rising at one point more than 13,000 feet above a depressed floor. It presents some very suggestive examples of terraces and large depressions, the latter especially well seen on the S.E. The bright interior includes a large central mountain with a digitated base on the S.E., some smaller hills on the S. of it, a deep crater W. of it (with small craters N. and S.), and, between the crater and the foot of the W. wall, a very curious winding cleft.

LASSELL.--This ring-plain, some 14 miles in diameter, is irregular both as regards its outline and the width of its rampart. There is a crater on the crest of the N.W. wall, just above a notable break in its continuity through which a ridge from the N.W. passes. There is another crater on the opposite side. The central mountain is small and difficult to see. About 20 miles N.E. of Lassell is a remarkable mountain group associated with a bright crater, and further on in the same direction is a light oval area, about 10 miles across, with a crater (Alpetragius d) on its S. edge. Madler described this area as a bright crater, 5 miles in diameter, which now it certainly is not.

LALANDE.--A very deep ring-plain, about 14 miles in diameter, N.E. of Ptolemaeus, with bright terraced walls, some 6000 feet above the floor, which contains a low central mountain. On the N. is the long cleft running, with some interruptions, in a W.N.W. direction towards Reaumur.

DAVY.--A deep irregular ring-plain, 23 miles across, on the Mare E. of Alphonsus. There is a deep crater with a bright rim on its S.W. wall, and E. of this a notable gap. There is also a wide opening on the N. The E. wall is of the linear type. A cleft crosses the interior.

GUERIKE.--The most southerly member of a remarkable group of partially destroyed walled-plains, standing in an isolated position in the Mare Nubium. Its border, on the W. and N. especially, is much broken, and never rises much more than 2000 feet above the Mare, except at one place on the N., where there is a mountain about 1000 feet higher. The E. wall is tolerably continuous, but is of a very abnormal shape. On the S. there is a peculiar LAMBDA-shaped gap (with a bright crater, and another less prominent on the E. side of it), the narrowest part of which opens into a long wide winding valley, bounded by low hills, extending to the W. side of a bright ring-plain, Guerike B, on the S.E. A crater-chain occupies the centre of this valley. There is much detail within Guerike. A large deep bright crater stands under the E. border on a mound, which, gradually narrowing in width, extends to the N. wall; and a rill-like valley runs from the N. border towards the E. side of the LAMBDA-shaped gap. In addition to these features, there is a shallow rimmed crater, about midway between the extremities of the rill-valley, and several minor elevations on the floor.

On the broken N. flank of Guerike is a number of incomplete little rings, all open to the N.; and E. of these commences a linear group of lofty isolated mountain masses extending towards the W. side of Parry, and prolonged for 30 miles or more towards the north. They are arranged in parallel rows, and remind one of a Druidical avenue of gigantic monoliths viewed from above. They terminate on the S. side of a large bright incomplete ring (with a lofty W. wall), connected with the W. side of Parry.

PARRY.--A more complete formation than Guerike. It is about 25 miles in diameter, and is encompassed by a bright border, which, at a point on the E., is nearly 5000 feet in height. It is intersected on the N. by passes communicating with the interior of Fra Mauro. There is a crater, nearly central, on the dusky interior, which, under a low sun, when the shadows of the serrated crest of the W. wall reach about half-way across the floor, appears to be the centre of three or four concentric ridges, which at this phase are traceable on the E. side of it. There is a conspicuous crater on the E. wall, below which originates a distinct cleft. This object skirts the inner foot of the E. border, and after traversing the N. wall, strikes across the wide expanse of Fra Mauro, and is ultimately lost in the region N. of this formation. Parry A, S. of Parry, is a very deep brilliant crater with a central hill and surrounded by a glistening halo. A cleft, originating at a mountain arm connected with the E. side of Guerike, runs to the S. flank of this object, and is probably connected with that which skirts the floor of Parry on the E.

BONPLAND.--A ruined walled-plain with a low and much broken wall, which on the S.W. appears to be an attenuated prolongation of that of Parry. It is of the linear type, the formation approximating in shape to that of a pentagon. The floor is crossed from N. to S. by a fine cleft which originates at a crater beyond the S. wall, and is visible as a light streak under a high light. Schmidt shows a short cleft on the W. of this.

FRA MAURO.--A large enclosure of irregular shape, at least 50 miles from side to side, abutting on Parry and Bonpland. In addition to the cleft which crosses it, the floor is traversed by a great number of ridges, and includes at least seven craters.

THEBIT.--A fine ring-plain, 32 miles in diameter, on the mountainous W. margin of the Mare Nubium, N.E. of Purbach. Its irregular rampart is prominently terraced, and its continuity on the N.E. interrupted by a large deep crater (Thebit A), at least 9 miles in diameter, which has in its turn a smaller crater, of about half this size, on its margin, and a small central mountain within, which was once considered a good optical test, though it is not a difficult object in a 4 inch achromatic, if it is looked for at a favourable phase. The border of Thebit rises at one place on the N.W. to a height of nearly 10,000 feet above the interior, which includes much detail. The E. wall of Thebit A attains the same height above its floor, which is depressed more than 5000 feet below the Mare.

BIRT.--This ring-plain, about 12 miles in diameter, is situated on the Mare Nubium, some distance due E. of Thebit. It has a brilliant border, surmounted by peaks rising more than 2000 feet above the Mare, and a very depressed floor, which does not appear to contain any visible detail. A bright crater adjoins it on the S.W., the wall of which at the point of junction is clearly very low, so that under oblique light the two interiors appear to communicate by a narrow pass or neck filled with shadow. I have frequently seen a break in the N.W. wall of Birt, which seems to indicate the presence of a crater. There is a noteworthy cleft on the E., which can be traced from the foot of the E. wall to the hills on the N.E. It is a fine telescopic object, and, under some conditions, the wider portion of it resembles a railway cutting traversing rising ground, seen from above. It is visible as a white line under a high light.

THE STRAIGHT WALL.--Sometimes called "the railroad," is a remarkable and almost unique formation on the W. side of Birt, extending for about 65 miles from N.E. to S.W. in a nearly straight line, terminating on the south at a very peculiar mountain group, the shape of which has been compared to a stag's horn, but which perhaps more closely resembles a sword-handle,--the wall representing the blade. When examined under suitable conditions, the latter is seen to be slightly curved, the S. half bending to the west, and the remainder the opposite way. The formation is not a ridge, but is clearly due to a sudden change in the level of the surface, and thus has the outward characteristics of a "fault" Along the upper edge of this gigantic cliff (which, though measures differ, cannot be anywhere much less than 500 feet high) I have seen at different times many small craterlets and mounds. Near its N. end is a large crater, and on the W. is a row of hillocks, running at right angles to the cliff. No observer should fail to examine the wall under a setting sun when the nearly perpendicular E. face of the cliff is brilliantly illuminated.

NICOLLET.--A conspicuous little ring-plain on the E. of Birt, and somewhat smaller. Between the two is a still smaller crater, from near which runs a low mountain range, nearly parallel to the straight wall, to the region S.E. of the Stag's Horn Mountains. Here will be found three small light-surrounded craters arranged in a triangle, with a somewhat larger crater in the middle.

PURBACH.--An immense enclosure of irregular shape, approximating to that of a rhomboid with slightly curved sides. It is fully 60 miles across, and the walls in places exceed 8000 feet in altitude, and include many depressions, large and small. On the E. inner slope are some fine terraces and several craters. The continuity of the circumvallation is broken on the N. by a great ring-plain, on the floor of which I have seen a prominent cleft and a crater near the S. side. There is a large bright crater in the interior of Purbach, S. of the centre, two others on the W. half of the floor, and a few ridges.

REGIOMONTANUS.--A still more irregular walled-plain, of about the same area, closely associated with the S. flank of Purbach, having a rampart of a similar complex type, traversed by passes, longitudinal valleys, and other depressions. Schmidt alone shows the especially fine example of a crater-row, which is not a difficult object, in connection with the S.E. wall. Excepting one crater, nearly central, and some inconspicuous ridges, I have seen no detail on the floor. Schmidt, however, records many features.

WALTER.--A great rhomboidal walled-plain, 100 miles in diameter, with a considerably depressed floor, enclosed by a rampart of a very complex kind, crowned by numerous peaks, one of which, on the W., rises 10,000 feet above the interior. If the formation is observed when it is close to the morning terminator, say, when the latter lies from I deg. to 2 deg. E. of the centre of the floor, it is one of the most striking and beautiful objects which the lunar observer can scrutinize. The inner slope of the border which abuts on Regiomontanus, examined at this phase under a high power, is seen to be pitted with an inconceivable number of minute craters; and the summit ridge, and the region towards Werner, scalloped in a very extraordinary way, the engrailing (to use an heraldic term) being due to the presence of a row of big depressions. The floor at this phase is sufficiently illuminated to disclose some of its most noteworthy features. Taking its area to be about 8000 square miles, at least 1200 square miles of it is occupied by the central mountain group and its adjuncts, the highest peak rising to a height of nearly 5000 feet (or nearly 600 feet higher than Ben Nevis), above the interior, and throwing a fine spire of shadow thereon. In the midst of this central boss are two deep craters, one being about 10 miles in diameter, and a number of shallower depressions. In association with the loftiest peak, I noted at 8 h., March 9, 1889, two brilliant little craters, which presumably are not far from the summit. Near the E. corner of the floor there is another large deep crater, and, ranging in a line from the centre to the S.E. wall, three smaller craters.

LEXELL.--On the E. of Walter extends an immense plain of irregular outline, which is at least equal to it in area. Though no large formation is found thereon; many ridges, short crater-rows, and ordinary craters figure on its rugged superficies; and on its borders stand some very noteworthy objects, among them, on the S., the walled-plain Lexell, about 32 miles in diameter, which presents many points of interest. Its irregular wall, rising, at one point on the S.W., to a height of nearly 8000 feet, is on the N.W. almost completely wanting, only very faint indications of its site being traceable, even under a low morning sun. On the opposite side it is boldly terraced, and has a large crater on its summit. The interior, the tone of which is conspicuously darker than that of the region outside, contains a small central hill, with two craters connected with it. The low N.W. margin is traversed by a delicate valley, which, originating on the N. side of the great plain, crosses the W. quarter of Lexell and terminates apparently on the S.W. side of the floor.

HELL.--A prominent ring-plain, about 18 miles in diameter, on the E. side of the great plain. There is a central mountain and many ridges within.

BALL.--A somewhat smaller ring-plain on the S.E. edge of the great plain, with a lofty terraced border and a central mountain more than 2000 feet high. There are two large irregular depressions on the W. of the formation, a crater on the S., and a smaller one on the N. wall.

PITATUS.--This remarkable object, 58 miles in diameter, with Hesiodus, its companion on the E., situated at the extreme S. end of the Mare Nubium, afford good examples of a class of formations which exhibit undoubted signs of partial destruction, from some unknown cause, on that side of them which faces the Mare. On every side but the N., Pitatus is a walled plain of an especially massive type, the border on the S.E. furnishing one of the finest examples of terraces to be found on the visible surface. On the S.W., two parallel rows of large crateriform depressions, perhaps the most remarkable of their kind, extend for 60 miles or more to the W. flank of Gauricus. On the N.W., the rampart includes many curious irregular depressions and craters, and gradually diminishes in height, till, for a space of about 12 miles on the N., there can hardly be said to be any border at all, its site being marked by some inconsiderable mounds and shallow hollows. There is a small bright central mountain on the floor, and, S. of it, two larger but lower elevations. A distinct straight cleft traverses the N.W. side of the interior very near the wall, to which it forms an apparent chord, and a second cleft occupies a similar position with respect to the bright N.E. border. A narrow pass forms a communication with the interior of Hesiodus.

HESIODUS.--This walled-plain, little more than half the diameter of the last, has an irregular outline, and for the most part linear walls, which on the S. are massive and lofty (4000 feet), but on the N. very low, and broken by gaps. There is a fine deep crater on the S. border, and a small but distinct crater on the floor, nearly central, the only object thereon which I have seen, though Schmidt draws a smaller one on the W. of it.

A mountain abutting on the N.E. side of Hesiodus is the W. origin of one of the longest clefts on the moon. Running in an E.S.E. direction, it traverses the Mare to a crater near the W. face of the Cichus mountain arm, reappears on the E. side of this object, and is finally lost amid the hills on the N. of Capuanus. The W. section of this cleft is coarser and much more distinct than that lying E. of the mountain arm.

GAURICUS.--A large walled-plain S. of Pitatus, about 40 miles in diameter. The border is very irregular, and, according to Neison, consists on the E. of a precipitous cliff more than 9000 feet high. It is surrounded by a number of large rings on the S., and has several considerable small depressions on its N. border. There is apparently no prominent detail on the floor. Schmidt shows some ridges and craterlets.

WURZELBAUER.--Another irregular walled-plain, about 50 miles in diameter, on the S.E. of Pitatus, with a very complex border, in connection with which, on the S.W., is a group of fine depressions, and on the S.E. a large crater. There is much detail on the very uneven floor.

MILLER.--One of a group of three moderately large ring-plains, of which Nasireddin is a member, near the central meridian in S. latitude 39 deg. Its massive border rises nearly 11,000 feet above the floor, on which stands a central peak. Miller is about 36 miles in diameter.

NASIREDDIN.--A somewhat smaller ring-plain on the S. of the last, and of a very similar type. It contains a central peak and several minor elevations. Between its N.W. border and the S.W. flank of Miller is a smaller ring-plain of about half the size of Nasireddin, and on the S.E. a large enclosure named HUGGINS.

ORONTIUS.--Huggins has encroached on the W. side of this irregular ring-plain and overlaps it. It is of considerable size. The floor includes much detail and a prominent crater.

SASSERIDES.--A formation of irregular shape, with very lofty walls, situated amid the confusion of ring-plains, craters, crater-pits, &c., in the region N. of Tycho, some of which are fully as deserving of a distinct name.

HEINSIUS.--A very curious formation on the N.E. of Tycho: a fine telescopic object under oblique illumination. It has an irregular but continuous border, except on the S., where two large ring-plains have encroached upon it, and a third, N. of a line joining their centres, occupies no inconsiderable portion of the floor. Heinsius is nearly 50 miles across, and the border on the W., is nearly 9000 feet above the interior, which includes, at least, three small craters. The walls of the intrusive ring-plains have craters on their summits; the more westerly has two on the W., and its companion, one on the S.W. The ring-plain on the floor has a crater on its E. wall. Schmidt shows a small crater between the ring-plains on the S. border.

SAUSSURE.--A ring-plain W. of Tycho, 28 miles in diameter, with bright lofty terraced walls and a somewhat dark interior, on which there is a crater, W. of the centre, and some crater-pits. There are several large depressions on the S.W. wall. It is surrounded by formations which, though nearly as prominent as itself, have not, with the exception of Pictet on the E., and one on the N.W., called Huggins by Schmidt, received distinctive names. The region W. of Saussure abounds in craterlets, some of which are of the minutest type. One of the Tycho streaks is manifestly deflected from its course by this formation, and another is faintly traceable on the floor.

PICTET.--A walled-plain of irregular shape, about 30 miles across, between Saussure and Tycho, with a border broken on the S. by a large conspicuous ring-plain, which is at least 10 miles in diameter, and, according to Schmidt, has a central mountain. Schmidt draws the S.E. border of Pictet as broken by ridges extending on to the floor. He also shows several craters and minor elevations thereon.

TYCHO.--As the centre from which the principal bright ray-system of the moon radiates, and the most conspicuous object in the southern hemisphere, this noble ring-plain may justly claim the pre-eminent title of "the Metropolitan crater." It is more than 54 miles in diameter, and its massive border, everywhere traversed by terraces and variegated by depressions within and without, is surmounted by peaks rising both on the E. and W. to a height of about 17,000 feet above the bright interior, on which stands a magnificent central mountain at least 5000 feet in altitude. Were it not somewhat foreshortened, Tycho would be seen to deviate considerably from what is deemed to be the normal shape. On the S. and W. especially, the wall approximates to the linear type, no signs of curvature being apparent where these sections meet. The crest on the S. and S.E. exhibits many breaks and irregularities; and it is through a narrow gap on the S. that a rill-like valley, originating at a small depression near the foot of the S.W. glacis, passes, and, descending the inner slope of the S.E. wall obliquely, terminates near its foot.

There is a distinct crater on the summit ridge on the S.E., and another below the crest on the outer S.W. slope. On the S. inner slope I have often remarked a number of bright oval objects, which, for the lack of a better word, may be termed "mounds" though they represent masses of material many miles in length and breadth. The outer slope of Tycho, exhibiting under a high light a grey nimbus encircling the wall, includes--craters, crater-pits, shallow valleys, spurs and buttresses--in short, almost every variety of lunar feature is represented. Excepting the central mountain and a crater on the W. of it, I have not seen any object on the floor, which, for some unexplained reason, is never very distinct. Schmidt shows several low ridges on the N.E. side. In a paper recently published in the Astronomische Nachrichten, Professor W.H. Pickering, describing his observations of the Tycho streaks made at Arequipa, Peru, with a 13 inch achromatic, asserts that they do not radiate from the centre of Tycho, but from a multitude of minute craters on its S.E. or N. rim. (See Introduction.)

MAGINUS.--An immense partially ruined enclosure, at least 100 miles from side to side, on the S.W. of Tycho, from which it is separated by a region covered with a confused mass of ring-plains and craters. On almost every part of its broken border stand large ring-plains, many of which, if they were isolated, or situated in a less disturbed region, would rank as objects of importance; but among such a multitude of features they pass unnoticed. The largest of them occupies no inconsiderable part of the S.E. wall, and is quite 30 miles in diameter, its own border being also much broken by depressions, as, indeed, are those of almost all the six or more large ring-plains which define the N. limits of Maginus. The loftiest portion of what remains of a true border rises at one place to more than 14,000 feet. On the floor, which is traversed by some of the Tycho rays, there is a mountain group associated with a crater, nearly central, and several large rings on the N. side. Though the formation is very difficult to detect under a high sun, Madler's dictum that "the full moon knows no Maginus" is not strictly true.

STREET.--A walled-plain between Tycho and Maginus, about 28 miles in diameter, with a border of moderate height, broken by depressions on the N. There are some small craters and ridges within; but the surrounding region, with its almost endless variety of abnormally shaped formations, is far more worthy of the observer's attention.

DELUC.--The largest and most prominent member of a curious group of ring-plains on the S.W. of Maginus. It is about 28 miles in diameter, and is encircled by a wall some 7000 feet above the interior, which includes a crater. A large ring with a central mountain encroaches on the N. wall, and a smaller object of the same class on the S. wall.

CLAVIUS.--There are few lunar observers who have not devoted more or less attention to this beautiful formation, one of the most striking of telescopic objects. However familiar we may consider ourselves to be with its features, there is always something fresh to note and to admire as often as we examine its apparently inexhaustible details. It is 142 miles from side to side, and includes an area of at least 16,000 square miles within its irregular circumvallation, which is only comparatively slightly elevated above the bright plateau on the W., though it stands at least 12,000 feet above the depressed floor. At a point on the S.W. a peak rises nearly 17,000 feet above the interior, while on the E. the cliffs are almost as lofty. There are two remarkable ring-plains, each about 25 miles in diameter, associated, one with the N., and the other with the S. wall, the floors of both abounding in detail. The latter, however, is the most noteworthy on account of the curious corrugations visible soon after sunrise on the outer N. slope of its wall, resembling the ribbed flanks of some of the Java volcanoes. There are five large craters on the floor of Clavius, following a curve convex to the N., and diminishing in size from W. to E. The most westerly stands nearly midway between the two large ring-plains on the walls, the second (about two-thirds its area) is associated with a complex group of hills and smaller craters. Both these objects have central mountains. In addition to this prominent chain, there are innumerable craters of a smaller type on the floor, but they are more plentiful on the S. half than elsewhere. On the S.E. wall are three very large depressions. On the broad massive N.E. border, the bright summit ridge and the many transverse valleys running down from it to the floor, are especially interesting features. There are very clear indications of "faulting" on a vast scale where this broad section of the wall abuts on the N. side of the formation.

CYSATUS.--A regular walled-plain, apparently about 28 miles in diameter, forming the most northerly member of a chain of formations, of which Newton, Short, and Moretus, extending towards the S. limb, form a part. Its border rises nearly 13,000 feet above a floor devoid of prominent detail.

GRUEMBERGER.--A much larger and more irregular ring-plain, nearly 40 miles from wall to wall, on the E. side of Cysatus. Its W. border rises nearly 14,000 feet above the interior, which includes an abnormally deep crater, the bottom of which is 20,000 feet below the crest of the W. wall, and several small depressions and ridges. The inner E. slope is finely terraced.

MORETUS.--A magnificent object, 78 miles in diameter, but foreshortened into a flat ellipse. Its beautifully terraced walls and magnificent central mountain, nearly 7000 feet high, are very conspicuous under suitable conditions. The rampart on the E. is more than 15,000 feet above the floor, while on the opposite side it is about 5000 feet lower.

SHORT.--A fine but foreshortened ring-plain of oblong shape, squeezed in between Moretus and Newton. It is about 30 miles in diameter, and on the S.E., where its border and that of Newton are in common, it rises nearly 17,000 feet above the interior, which includes, according to Neison, a small central hill. Schmidt shows a crater on the N. side of the floor.

NEWTON.--Is situated on the S.E. side of Short, and is the deepest walled-plain on the visible surface. It is of irregular form and about 143 miles in extreme length. One gigantic peak on the E. rises to nearly 24,000 feet above the floor, the greater part of which is always immersed in shadow, so that neither the earth or sun can at any time be seen from it.

MALAPERT.--A ring-plain situated far too near the limb for useful observation. Between it and Newton is a number of abnormally shaped enclosures.

CABEUS.--Another object out of the range of satisfactory scrutiny. Madler considered that it is as deep as Newton. According to Neison, a central peak and two craters can be seen within under favourable conditions. Schmidt draws a long row of great depressions on the N. side of it.

EAST LONGITUDE 20 deg. TO 40 deg.

LANDSBERG.--A ring-plain, about 28 miles in diameter, situated in Mare Nubium, S.E. of Reinhold, which in many respects it resembles. Its regular massive border is everywhere continuous. Only a solitary crater breaks the uniformity of its crest, that rises on the W. to nearly 10,000 feet, and on the E. to about 7000 feet above the floor, which is depressed about 7000 feet below the surrounding surface. The inner slopes exhibit some fine terraces, and on the broad W. glacis is a curious winding valley, which runs up the slope from the S.W. side to the crater just mentioned, then, bending downwards, joins the plain at the foot of the N. wall. Neither this nor the crater is shown in the maps. The large compound central peak is apparently the sole object in the interior. At 8 h. 25 m. on January 23, 1888, when observing the progress of sunrise on this formation with a 8 1/2 inch Calver-reflector charged with different eyepieces, I noticed, when about three-fourths of the floor was in shadow, that the illuminated portion of it was of a dark chocolate hue, strongly contrasting with the grey tone of the surrounding district. This appearance lasted till the interior was more than half illuminated, gradually becoming less pronounced as the sun rose higher on the ring. E. and S.E. of Landsberg is a number of ring-plains and craters well worthy of careful examination. Five of the largest are surrounded by a glistening halo, and one (that nearest to the formation) and another (the largest of the group) have each a minute crater on their N. wall.

EUCLIDES.--One of the most brilliant objects on the moon; a crater 7 miles in diameter, standing on a large bright area in the Mare Procellarum, E. of the Riphaean Mountains. Its E. rim rises nearly 2000 feet above the bright depressed floor; on the W. there is a bright little unrecorded crater.

WICHMANN.--This bright crater, about 5 miles in diameter, stands on a light area in Oceanus Procellarum, N.N.W. of Letronne and nearly due E. of Euclides. Some distance on the N.E. are the relics of what appears once to have been a large enclosure, represented now by a few isolated mountains.

HERIGONIUS.--A ring-plain, about 7 miles in diameter, in the Mare Procellarum, N.W. of Gassendi. There is a small crater a few miles S.E. of it, among the bright little mountains which flank this formation. Herigonius has a small central mountain, which is a good test for moderate apertures.

GASSENDI.--One of the most beautiful telescopic objects on the moon's visible surface, and structurally one of the most interesting and suggestive. It is a walled-plain, 55 miles in diameter, of a distinctly polygonal type, the N.W.

and S.W. sections being practically straight, while the intermediate W. section exhibits a slightly convex curvature, or bulging in towards the interior. There is also much angularity about the E. side, which is evident at an early stage of sunrise. The wall on the N. is broken through and almost completely wrecked by the great ring- plain Gassendi A. The bright eastern section of the border is in places very lofty, rising at one peak, N. of the well-known triangular depression upon it, to 9000 feet, and at other peaks on the same side still higher. It is very low on the S., being only about 500 feet above the surface. The floor, however, on the N. stands 2000 feet above the Mare Humorum. On the W. there is a peak towering 4000 feet above the wall, which is here about 5000 feet above the floor, and 8000 feet above the Mare Nubium. A very notable feature in connection with this formation is the little bright plain bounding it on the N.W., and separated from it by merely a narrow strip of wall. This enclosure is flanked on the N.E. by Gassendi A, and on the S.W. and N.W. by a coarse winding ridge, running from the W. wall and terminating at a large irregular dusky depression. Gaudibert has detected a crater near the S.E. edge of this bright plain, which includes also some oval mounds. The interior of Gassendi is without question unrivalled for the variety of its details, and, after Plato, has perhaps received more attention from observers than any other object. The bright central mountain, or rather mountains, for it consists of a number of grouped masses crowned by peaks, of which the loftiest is about 4000 feet, is one of the finest on the moon. It was carefully studied with a 6 1/4 inch Cooke-achromatic by the late Professor Phillips, the geologist, who compared it to the dolomitic or trachytic mountains of the earth. The buttresses and spurs which it throws out give its base a digitated outline, easily seen under suitable illumination. There are between 30 and 40 clefts in the interior, the majority being confined to the S.W. quarter of the floor. Those most easily seen pertain to the group which radiates from the central mountain towards the S.W. wall. They are all more or less difficult objects, requiring exceptionally favourable weather and high powers. A fine mountain range, the Percy Mountains, is connected with the E. flank of Gassendi, extending in a S.E. direction towards Mersenius, and defining the N.E. side of the Mare Humorum.

BULLIALDUS.--A noble object, 38 miles in diameter, forming with its surroundings by far the most notable formation on the surface of the Mare Nubium, and one of the most characteristic ring-plains on the moon. It should be observed about the time when the morning terminator lies on the W.

border of the Mare Humorum, as at this phase the best view is obtained of the two deep parallel terrace valleys which run round the bright inner slope of the E. wall, of the crater-row against which they abut on the S.E., and of the massive W. glacis, with its spurs and depressions. The S. border of Bullialdus has been manifestly modified by the presence of the great ring-plain A, a deep irregular formation with linear walls, which is connected with it by a shallow valley. The rampart of Bullialdus rises about 8000 feet above a concave floor, which sinks some 4000 feet below the Mare on the E. With the exception of the fine compound central mountain, 3000 feet high, there are few details in the interior. On the S., is the fine ring-plain B, connected with the S.E. wall near the crater-row by a well-marked valley, and nearly due E. of B is another, a square-shaped enclosure, C, with a very lofty little mountain on the E. side of it, and a crater on its S. wall. In addition to these features, there are many ridges and surface inequalities, very prominent under oblique illumination.

LUBINIEZKY.--A regular enclosure, about 23 miles in diameter, N.E. of Bullialdus, with a low attenuated border, which is nowhere more than 1000 feet in height. It is tolerably continuous, except on the S., where there are two or three breaks. Its level dark interior presents no details to vary its monotony. Close under the N.W. wall is a small crater connected with it by a ridge, and E. of this a very rugged area, traversed in every direction by narrow shallow valleys, which are well worth looking at when close to the morning terminator. A bright spur projects from the N. wall of Lubiniezky.

KIES.--A somewhat similar formation, S. of Bullialdus, about 25 miles in diameter, also encircled by a border of insignificant dimensions, attaining an altitude of 2400 feet at only one point on the S.E., while elsewhere it is scarcely higher than that of Lubiniezky. It is clearly polygonal, approximating to the hexagonal type. On the more distinct S. section a bright spur projects from it. On the N. its continuity is broken by a distinct little crater. It is traversed by a remarkable white streak, extending in a S.W. direction from Bullialdus C (where it is very wide), across the interior, to the more westerly of two craters S.W. of Mercator. Another streak branches out from it near the centre of the floor, and runs to the W. wall. The principal streak, so far as the portion within Kies is concerned, represents a cleft. On the Mare E. of Kies is a curious circular mound, and farther towards Campanus two prominent little mountains. On the N.W. is a large obscure ring and a wide shallow valley bordered by ridges.

AGATHARCHIDES.--A very irregular complex ring-plain, about 28 miles in diameter, forming part of the N.W. side of the Mare Humorum. It must be observed under many phases before one can clearly comprehend its distinctive features. The wall is very deficient on the N., but is represented in places by bright mountain masses. The formation is flanked on the E. by a double rampart, which is at one place more than 5000 feet in height, with a deep intervening valley. The S. wall is traversed by a number of parallel valleys, all trending towards Hippalus. These are included in a much wider and longer chasm, which, gradually diminishing in breadth, extends up to the N. wall of the latter.

HIPPALUS.--A partially ruined walled-plain, about 38 miles in diameter, on the W. side of the Mare Humorum, S. of Agatharchides. At least one-third of the border is wanting on the S.E., but under a low sun its site can be distinguished by a faint marking and the obvious difference in tone between the dark interior and the lighter-coloured plain. The rest of the wall is bright and continuous, except at a place on the W., where what appears to be the segment of a large ring has encroached upon it. There are two craters in the interior of Hippalus, and a row of parallel ridges, running obliquely from the S.W. wall up to a cleft which traverses the floor from N. to S. W. of Hippalus stands a bright crater, Hippalus A, with an incomplete little ring-plain adjoining it on the N.W.; and N.E. of it a much larger obscure ring containing two little hills. The Hippalus rill-system is a very interesting one, and the greater part of it can, moreover, be easily traced in a good 4 inch achromatic. It originates in the rugged region E. of Campanus, from which five nearly parallel curved clefts extend up to the rocky barrier, connecting the N. side of this formation with the S.W. side of Hippalus. The most westerly of these furrows is interrupted by a crater on this wall, but reappears on the N. side of it, and, after making a detour towards the W. to avoid a little mountain in its path, runs partially round the E. flank of Hippalus A, and then, continuing its northerly course, terminates amid the mountains W. of Agatharchides. (A short parallel cleft runs E. of this from the little mountain to the E. side of A.) The most easterly member of the system, originating N. of Ramsden, enters Hippalus at the S. side of the great gap in the border, and, after traversing the floor at the W. foot of a ridge thereon, also extends towards the mountains W. of Agatharchides. Between these clefts are three intermediate furrows, one of which runs N. from the N. side of the encroaching ring already referred to, on the W. wall of Hippalus.

CAMPANUS.--A ring-plain, 30 miles in diameter, on the rocky barrier, extending in nearly a straight line from Hippalus to Cichus. Its terraced walls, which rise on the E. more than 6000 feet above the floor, are broken on the S. by a narrow valley, and on the E. by a small crater. A small central mountain is apparently the only object on a very dark interior.

MERCATOR.--A more irregular ring-plain of about the same area, adjoining Campanus on the S.W. Its rampart is somewhat lower, and is partially broken on the N. by two semi-rings, and on the S. by a gap. The E. wall extends on the S. far beyond the limits of the formation, and terminates in a brilliant mountain mass 6000 feet in height. There is a bright crater on the crest of both the E. and W. border. On the plain E. of Mercator is a remarkable little crater standing on a light area, and, just under the wall, a dusky pit connected with it by a rill-like marking. These objects are of a very doubtful nature, and should be carefully observed. The floor of Mercator is much lighter than that of Campanus, and appears to be devoid of detail.

CICHUS.--A conspicuous ring-plain, about 20 miles in diameter, with a prominent deep crater about 6 miles across on its E. rim. It is situated on a curious boot-shaped plateau, near the S. end of the rocky mountain barrier associated with the last two formations. Its walls rise about 9000 feet above a sunken floor, on which there is some faint detail, but apparently nothing deserving the distinction of a central mountain. The plateau on the N. is cut through by a fine broad valley, which has obviously interfered with a large crateriform depression on its southern edge. A cleft runs from a small crater W. of the plateau up to this valley, and extends beyond to the W. wall of Capuanus. There is also a delicate cleft crossing the region S. of Cichus to the group of complicated formations S.W. of Capuanus. As already mentioned, the great Hesiodus cleft is associated with the Cichus plateau.

CAPUANUS.--A large ring-plain, about 34 miles in diameter, E. of Cichus, with a border especially remarkable on the E., where it rises more than 8000 feet above the outside country, and includes a large brilliant shallow crater. It is broken on the N.W. by a small but noteworthy double crater; and on the S. its continuity is destroyed for many miles by a number of big circular and sub-circular depressions and prominent deep valleys, far too numerous and complicated to describe. The level dusky interior contains only a low mound on the S., but is crossed by some light streaks running from N. to S.

RAMSDEN.--This ring-plain, 12 miles in diameter, derives its importance from the remarkable rill-system with which it is so closely associated. Its border, about 1800 feet on the W. above the outside surface, is slightly terraced within on the E., where there is an unrecorded bright crater on the slope. The two principal clefts on the S. originate among the hills E. of Capuanus. The more easterly begins at a crater on the N. edge of these objects, and runs N. to the E. side of Ramsden; the other originates at a larger crater, and proceeds in a N. direction up to a bright little mountain S.W. of Ramsden; when, swerving to the N.E., it ends at the W. wall of this formation. This mountain is a centre or node from which three other more delicate branches radiate. On the N., three of the shortest clefts pertaining to the system are easily traceable from neighbouring mountains up to the N. wall, which they apparently partially cut through. The E. pair have a common origin, but open out as they approach the border of Ramsden.

VITELLO.--A very peculiar ring-plain, 28 miles in diameter, on the S. side of the Mare Humorum, remarkable for having another nearly concentric ring-plain, of considerably less altitude within it, and a large bright central boss, overlooking the inner wall, 1700 feet in height. The outer wall is somewhat irregular, and is broken by gaps and valleys on the S. and N.W. It rises on the E. about 5000 feet above the Mare, but only about 2000 above the interior, which includes a crater on its N. side, and some low ridges.

HAINZEL.--This remarkable formation, which is about 55 miles in greatest length, but is hardly half so broad, derives its abnormal shape from the partial coalescence of two nearly equal ring-plains, the walls of both being very lofty,--more than 10,000 feet. It ought to be observed under a morning sun when the floor is about half illuminated. At this phase the extension of the broad bright terraced E. border across a portion of the interior is very apparent, and the true structural character of the formation clearly revealed. The floor abounds in detail, among which, on the S., are some large craters and a bright longitudinal ridge. Hainzel is flanked on the W. and S.W. by a broad plateau, W. of which stand two ring-plains about 15 miles in diameter, both having prominent central mountains and bright interiors.

WILHELM I.--A large irregular formation, about 50 miles across, S.E. of Heinsius, with walls varying very considerably in height, rising more than 11,000 feet on the E., but only about 7000 feet on the opposite side. The border is everywhere crowded with depressions, large and small. Three ring-plains, not less than 6 miles in diameter, stand upon the S. wall, the most westerly overlapping its shallower neighbour on the E., which projects beyond the wall on to the floor. The interior has a very rugged and uneven surface, upon the N. side of which are two very distinct craters, and a short crater-row on the W. of them. It is traversed from W. to E. by three bright streaks from Tycho, two on the N. being very prominent under a high light.

LONGOMONTANUS.--A much larger walled-plain, S. of the last. It is 90 miles in diameter, with a border much broken by depressions, especially on the N.E. At one peak on this side it rises to the tremendous altitude of 13,000 feet above the floor, and at peaks on the W. more than 1000 feet higher. There is a crowd of ring-plains on the N.E. quarter of the interior, and some hills and craterlets in other parts of it. It is also crossed by rays from Tycho.

SCHILLER.--A fine lozenge-shaped enclosure, with a continuous but somewhat irregular border. It is about 112 miles in extreme length, and rather more than half this in breadth. The loftiest section of the wall is on the W., where it rises 13,000 feet above a considerably depressed interior. There is a bright crater on this side and some terraces. On the broad inner slope of the E. border, the summit ridge of which is especially well-marked, there is a large shallow depression. The floor contains scarcely any detail, except some ridges on the N. side and a few craterlets. The great bright plain E. of Schiller and the region on the S.E. are especially worthy of scrutiny under a low morning sun.

BAYER.--This object, 29 miles in diameter, with a terraced border rising on the W. to a height of 8000 feet above the floor, is so closely associated with Schiller, that it may almost be regarded as forming part of it. A long lofty mountain arm, apparently connected with the W. wall of the latter, runs from the E. side of Bayer towards the N.W. There is a crater on the E. side of the interior.

ROST.--An oblong-shaped ring-plain, 30 miles in diameter, on the S.W. of

Schiller, with moderately high walls, and, according to Neison, a shallow depression within, nearly central. I have seen a crater shown by Schmidt on the E. side of the floor. A valley runs from the E. side of Rost to the S. of Schiller.

WEIGEL.--A not very conspicuous ring-plain on the S. of Schiller, with a crater on its N.W. rim, and a larger ring adjoining it on the S.E. A prominent curved mountain arm from the E. wall of Schiller runs towards the N. side of this formation.

BLANCANUS.--A formation, 50 miles in diameter, on the S.E. side of Clavius, whose surpassing beauties tend to render the less remarkable features of this magnificent ring-plain and those of its neighbour Scheiner less attractive than they otherwise would be. The crest of its finely terraced wall, which at one peak on the E. rises to 18,000 feet, is at least 12,000 feet above the interior. Krieger saw twenty craters on the floor (1894, Sept. 21, 13h.), most of them situated on the S. quarter.

SCHEINER.--A still larger object, being nearly 70 miles in diameter, with a prominently terraced wall, fully as lofty as that of Blancanus. There is a large crater, nearly central, two others on the N.E. side of the floor, and a fourth at the inner foot of the E. wall. There is also a shallow ring on the N.E. slope. Schmidt shows, but far too prominently, two straight ridges crossing each other on the S. side of the central crater.

CASATUS.--A large walled-plain, about 50 miles in diameter, S.E. of Blancanus, near the limb, remarkable for having one of the loftiest ramparts of all known lunar objects; it rises at one peak on the S.W. to the great height of 22,285 feet above the floor, while there are other peaks nearly as high on the N. and S. The wall is broken on the E. by a fine crater. There is also a crater on the N.W. side of the very depressed floor, together with some craterlets.

KLAPROTH.--Casatus partially overlaps this still larger but less massive formation on its S.E. flank. The walls of Klaproth are much lower and very irregular and broken, especially on the W. There are some ridges on the floor. The neighbouring region is covered with unnamed objects, large and small.

EAST LONGITUDE 40 deg. TO 60 deg.

FLAMSTEED.--A bright ring-plain, 9 miles in diameter, in a barren region in the Oceanus Procellarum, N.E. of Wichmann. It has a regular border (broken at one place on the N. by a gap, which probably represents a crater), rising to a height of about 1400 feet above the surrounding plain. A great enclosure, 60 miles in diameter, lies on the N. of Flamsteed. It is defined by low ridges which exhibit many breaks, though under a high light the ring is apparently continuous. Within are several small craters and two considerable hills, nearly central.

HERMANN.--A ring-plain, about 10 miles in diameter, in the Oceanus Procellarum, W. of Lohrmann. It is associated with a group of long ridges, running in a meridional direction and roughly parallel to the coast-line.

LETRONNE.--A magnificent bay or inflexion in the coast-line of the Oceanus Procellarum, N.N.E. of Gassendi, presenting an opening towards the N. of nearly 50 miles, and bounded on the S. and S.W. by the lofty Gassendi highlands. Its border on the W., about 3000 feet high, is crowned with innumerable small depressions. The interior includes four bright little mountains, nearly central (three of them forming a triangle), a bright crater on the W. side, and several minor elevations and ridges. On the plain N. of the bay, is a large bright crater, from which a fine curved ridge runs to the central mountains. If Letronne is observed under oblique illumination, the low mounds and ridges on the Mare outside impress one with the idea that they represent the remains of a once complete N. wall.

BILLY.--A ring-plain, 31 miles in diameter, S.E. of Letronne, with a very dark floor, depressed about 1000 feet below the grey surface on the W., and a regular border, rising more than 3000 feet above it. There is a narrow gap on the S., and indications of a crater on the N.W. rim. Two small craters stand on the S. half of the interior. The formation is flanked on the S.W. by highlands.

HANSTEEN.--A somewhat larger ring-plain, with a lower and more irregular rampart, rising on the W. to nearly 3000 feet above the floor, which is depressed to about the same extent as that of Billy. Both the inner and outer slopes are terraced on the E., where the glacis is traversed by a short, delicate, rill-like valley. There are some bright curved ridges on the floor. On the W. of

Billy and Hansteen is a wide inlet of the Oceanus Procellarum, bounded by the Letronne region on the W., and on the S. by lofty highlands. On the surface, not far from the S.W. border of Hansteen, is a curious triangular-shaped mountain mass, with a digitated outline on the S., and including a small bright crater on its area. Between this and the ring- plain is a large but somewhat obscure depression, N. of which lies a rill-like object extending from the N. point of the triangular mountain to the W. wall. At the bottom of a gently sloping valley between Billy and Hansteen is a delicate marking, which seems to represent a cleft connecting the two formations.

ZUPUS.--A formation about 12 miles in diameter with a dark floor, situated in the hilly region N.E. of Mersenius.

FONTANA.--A noteworthy ring-plain, about 20 miles in diameter, E.N.E. of Zupus, with a bright border, exhibiting a narrow gap on the S. and two large contiguous craters on the N.W. The faint central mountain stands on a dusky interior. On the N. is a large peculiar depressed plain with a gently sloping wall, within which are three short rill-like valleys and a crater.

MERSENIUS.--With its extensive rill-system and interesting surroundings, one of the most notable ring-plains in the third quadrant. It is 41 miles in diameter, and is encircled by a fine rampart, which on the side fronting the Mare Humorum rises 7000 feet above the floor, which is distinctly convex, and is depressed 3000 feet below the region on the E., though it stands considerably above the level of the Mare. The prominently terraced border is tolerably regular on the N.W., but on the S. and S.E. is much broken by craters and depressions, the largest and most conspicuous interrupting the continuity of its summit-ridge on the latter side. A fine crater-row traverses the central part of the interior, nearly axially, and a delicate cleft crosses the N. half of the floor from the inner foot of the N.E. wall to a crater not far from the opposite side. I detected another cleft on November 11, 1883, also crossing the N. side of the floor.

South of Mersenius is the fine ring-plain Mersenius d, about 20 miles in diameter, situated on the border of the Mare; and, extending in a line from this towards Vieta are two others (a, and Cavendish d,), somewhat larger, but otherwise similar; the more easterly being connected with Cavendish by a mountain arm. One of the principal clefts of the system (all of which run roughly parallel to the N.E. side of the Mare, and extend to the Percy Mountains E. of Gassendi) crosses the floor of d, and, I believe, partially cuts into its W. wall. Another, the coarsest, abuts on a mountain arm connecting d with Mersenius, and, reappearing on the E. side, runs up to the N.W. wall of the other ring- plain, a, and, again reappearing on the E. of this, strikes across the rugged ground between a and Cavendish d, traversing its floor and border, as does also another cleft to the N. of it. Cavendish d includes a coarse cleft on its floor, running from N. to S., which I have frequently glimpsed with a 4 inch achromatic. There are two other delicate clefts running from the Gassendi region to the S.W. side of Mersenius, which are in part crater-rills.

CAVENDISH.--A notable ring-plain, 32 miles in diameter, S.E. of Mersenius, with a prominently terraced border, rising at one point on the S. to a height of 6000 feet above the interior, on which are a few low ridges. A large bright ring-plain (e), about 12 miles in diameter, breaks the continuity of the S.E. wall, and adjoining this, but beyond the limits of the formation, is another smaller ring with a central hill. There is also a bright crater on the N.W. border. The W. glacis is very broad, and includes two large shallow depressions. An especially fine valley runs up to the N. wall, to the W. side of e.

VIETA.--One of the finest objects in the third quadrant; a ring-plain 51 miles in diameter, with broad lofty walls, a peak on the west rising to nearly 11,000 feet, and another N. of it to considerably more than 14,000 feet above the interior. It is bounded by a linear border, approximating very closely to an hexagonal shape, which is broken by many gaps and cross-valleys. On the S., the S.W. and S.E. sections of the wall do not meet, being separated by a wide valley flanked on the W. by a fine crater, which has broken down the rampart at this place. The N. border is likewise intersected by valleys and by a crater-row. The inner slopes are conspicuously terraced. There is a very inconspicuous central mountain and several large craters on the floor, some of them double. Ten have been counted on the N. half of the interior. On the S.E. of Vieta are two fine overlapping ring-plains, with a crater on the wall common to both.

DE VICO.--A conspicuous little ring-plain, about 9 miles in diameter, with a lofty border, some distance E. of Mersenius.

LEE.--An incomplete walled-plain, about 28 miles in diameter, on the S. side of the Mare Humorum, E. of Vitello, from which it is separated by another partial enclosure, with a striking valley, not shown in the published maps, running round its W. side. If viewed when its E. wall is on the morning terminator, some isolated relics of the wrecked N.W. wall of Lee are prominent, in the shape of a number of attenuated bright elevations separated by gaps. Within are three or four conspicuous hills.

DOPPELMAYER.--Under a high sun this large ring-plain, 40 miles in diameter, resembles a great bay open to the N.W., without a trace of detail to break the monotony of the surface on the side facing the Mare Humorum. When, however, it is viewed under oblique morning illumination, a low broad ridge is easily traceable, extending across the opening, indicating the site of a ruined wall. There is an isolated mountain at the S.W. end of this, which casts a fine spire of shadow across the floor at sunrise. The interior contains a massive bright central mountain and several little hills. The crest of the wall on the E. is much broken.

FOURIER.--A large ring-plain, 30 miles in diameter, S.W. of Vieta, with a border rising at a peak on the W. more than 9000 feet above the floor, There are two craters on the outer slope of the N.W. wall, a prominent crater on the S. wall, and (according to Schmidt) a small central crater on the floor, which I have not seen. In the region between Fourier and Vieta there are three ring-plains, two (the more westerly) standing side by side, and on the W., towards the Mare, are two others much larger, that nearer to Fourier being traversed by one cleft, and the other by two clefts, crossing near the centre of the floor.

CLAUSIUS.--A small bright ring-plain in an isolated position N.W. of Schickard, with a crater both on its N. and S. rim, and a faint central hill.

LACROIX.--A ring-plain 20 miles in diameter, N. of Schickard. It has a prominent central mountain.

SCHICKARD.--One of the largest wall-surrounded plains on the visible surface of the moon, extending about 134 miles from N. to S., and about the same from E. to W., enclosing a nearly level area, abounding in detail. Its border, to a great extent linear, is very irregular, and much broken by the interposition of small ring-plains and craters, and on the N. by cross-valleys. Its general height is about 4000 feet, the loftiest peak on the W. wall rising to more than 9000 feet above the floor. The inner slopes of this vast rampart are very complex, especially on the E., where many terraces and depressions may be seen under suitable illumination. There are three large ring-plains in the interior, all of them S. of the centre; and at least five smaller ones near the inner foot of the E. wall, which can only be well observed when libration is favourable. The two more easterly of the large ring-plains are connected by a cleft, and there are several short clefts and crater-rows associated with the smaller ring-plains. On the N. side of the area is a number of minute craters. The floor is diversified by two large dark markings--an oblong patch on the S.W. side, abutting on the wall, being the more remarkable; and a dusky area, occupying a great portion of the N. part of the floor, and extending up to the N. border. This is traversed by a light streak running from N. to S., which is the site of a row of minute craters.

LEHMANN.--A ring-plain, about 28 miles in length, on the N. of Schickard, with which it is connected by a number of cross-valleys.

DREBBEL.--A bright ring-plain, 18 miles in diameter, on the N.W. of Schickard, with a lofty irregular border (especially on the W.), exhibiting a well-marked terrace on the E., a distinct gap on the N., and a small crater on the S.E. rim. On a dusky area between it and Schickard stand three prominent deep craters.

PHOCYLIDES.--This extraordinary walled plain, with its neighbouring enclosures, is structurally very remarkable and suggestive. It consists of a large irregular formation, with a lofty wall, flanked on the N. by a smaller and still more irregular enclosure (b), the floor of which is 1500 feet above that of Phocylides, the line of partition being a high cliff, probably representing a "fault," whose shadow under a low sun is very striking. Phocylides is about 80 miles in maximum length, or, if we reckon the small enclosure b to form a part of it, more than 120 miles. The loftiest peak, nearly 9000 feet, is on the W. border, near the partition wall. The continuity of the rampart is broken on the S. by a large crater. There is a bright ring-plain on the W. side of the floor, and a few small craters. Phocylides b has only a solitary crater within it. Phocylides C, abutting on the W. flank of Phocylides, is about 26 miles in diameter. Its somewhat dusky interior is devoid of detail, but the outer slope of its W. wall is crowded with a number of minute craters, which, under good conditions, may be utilised as tests of the defining power of the telescope used. Phocylides A, on the bright S.W. plain, is a large deep crater with a fine crater-row flanking it on the W.

WARGENTIN.--A most remarkable member of the Phocylides group, flanking the S.E. side of Schickard. Unlike the majority of lunar formations, its floor is raised considerably above the surrounding region, so that it resembles a shallow oval dish turned upside down. It is 54 miles in diameter, and, except on the S.W. (where it abuts on Phocylides b, and for some distance is bounded by its wall), it has only a border of very moderate dimensions. On the N.E. slope of this ghostly rampart I have seen a distinct little crater, and two much larger depressions on the N.W. slope. There are some low ridges on the floor, radiating from a nearly central point, which have been aptly compared to a bird's foot.

SEGNER.--A fine ring-plain, 46 miles in diameter, on the S.E. side of Schiller, with a linear border on every side except the N. At a peak on the W., whose shadow is very remarkable, it rises to a height of more than 8000 feet above the outer surface. There is a crater on the S.W. wall, another on the N.W. wall, and several depressions on the outer slope on this side. The central mountain is small but conspicuous. A large unnamed enclosure extends N. of Segner: it is larger than Schiller, and is surrounded by a lofty barrier. The bright plain between this and the latter is worth examination under a low sun.

ZUCHIUS.--Is situated on the S.E. of Segner, which it slightly overlaps. It is very similar in size and general character, and has a lofty terraced wall, rising at one place on the W. to nearly 11,000 feet above the floor. A very fine chain of craters, well seen when the opposite border is on the morning terminator, runs round the outer W. slope of the wall. There is a bright crater beyond this on the S.W. Zuchius has a central peak.

BETTINUS.--Another ring-plain of the same type and size, some distance S. of the last, with a massive border, terraced within, and rising on the W. more than 13,000 feet above the floor, on which stands a grand central mountain, whose brilliant summit is in sunlight a long time before a ray reaches any part of the deep interior.

KIRCHER.--A ring-plain, about 45 miles in diameter, S. of Bettinus, remarkable also for its very lofty rampart, which on the S. attains the tremendous height of nearly 18,000 feet above the floor, which appears to be devoid of detail.

WILSON.--The most southerly of the chain of five massive ring-plains, extending in an almost unbroken line from Segner and differing only very slightly in size. It is about 40 miles in diameter, and has a somewhat irregular border, both as regards shape and height, rising at one peak on the S.W. to nearly 14,000 feet above a level interior, which apparently contains no conspicuous features.

EAST LONGITUDE 60 deg. TO 90 deg.

GRIMALDI.--This ranks among the largest wall-surrounded plains on the moon, and is perhaps the darkest. It extends 148 miles from N. to S. and 129 miles from E. to W., enclosing an area of some 14,000 square miles, or nearly double that of the principality of Wales. This vast dusky surface is bounded on the E. by a tolerably regular border, having an average height of about 4000 feet, while on the opposite side it is much broken, and in places considerably loftier, rising at one peak on the S.W. to an altitude of 9000 feet. About midway, also, this western rampart attains a great height, as may be seen by any one who observes at sunrise the magnificent shadow of it, and its many peaks thrown across the bluish-grey interior. On the S. the wall is broken by a large irregular depression, on the W. of which is a very curious V-shaped rill valley. On the N.W. it is comparatively low, and in places discontinuous; and even to a greater extent than on the S.W., intersected by passes. At the extreme N. end, a number of wide valleys cut through the wall and trend towards Lohrmann. There is a considerable ring-plain at the inner foot of the N.E. wall, but, except this and a few longitudinal ridges, just visible under a very low sun, there is apparently no other object to vary the monotony of this great expanse.

DAMOISEAU.--Consists of a complex arrangement of rings, an enclosure 23 miles in diameter, with a somewhat smaller enclosure placed excentrically within it (the N. side of both abutting on a bright plateau), with two large depressions intervening between their W. borders. This peculiarity, almost unique, renders the formation an especially interesting object. Damoiseau is situated on the W. side of Grimaldi, on the E. coast-line of the Oceanus Procellarum, from which the S.W. border rises at a gentle inclination. On the N.W. there is a curious curved inflexion of the Mare, bounded by a bright cliff, representing probably the E. side of a destroyed ring, a supposition which is strengthened by the existence of a faint scar on the surface of the sea, extending in a curve from one extremity of the bay to the other, and thus indicating the position of the remainder of the ring. A conspicuous little crater stands at the S. end of it, and two others some distance to the W. The smaller component of Damoiseau contains a low central ridge.

RICCIOLI.--An immense enclosure, near the limb, N.E. of Grimaldi, bounded by a rampart which is very irregular both in form and height, though nowhere of great altitude, and much broken by narrow gaps. It is especially low and attenuated on the N., where a number of ridges with intervening valleys traverse it. On the S. also a wide valley cuts through it. With the exception of a few low rounded hills and ridges, a short crater-row under the S.E. wall, and two small craters on the S.W., there are no details on the floor, which, however, is otherwise remarkable for the dusky tone of its surface, especially on the N. This dark patch occupies the whole of the N.E. side of the interior, and is bounded on the S. by an irregular outline, extending at one point nearly to the centre, and on the W. by a curved edge. The W. side is much darker than the rest. It is, in fact, as dark, if not darker, than any part of the floor of Grimaldi. Riccioli extends 106 miles from N. to S., and is nearly as broad. It includes an area of 9000 square miles.

ROCCA.--An irregular formation, 60 miles in length, near the limb S.E. of Grimaldi, consisting of a depression partially enclosed by mountain arms.

SIRSALIS.--The more westerly of a conspicuous pair of ring-plains about 20 miles in diameter, in the disturbed mountain region some distance S.W. of Grimaldi. It has lofty bright walls, rising to a great height above a depressed floor, on which there is a prominent central mountain. The E. border encroaches considerably on the somewhat larger companion, which is, however, scarcely a third so deep. One of the longest clefts on the visible surface runs immediately W. of this formation. Commencing at a minute crater on the N. of it, it grazes the foot of the W. glacis; then, passing a pair of small overlapping craters (resembling Sirsalis and its companion in miniature), it runs through a very rugged country to a ring-plain E. of De Vico (De Vico a), which it traverses, and, still following a southerly course, extends towards Byrgius, in the neighbourhood of which it is apparently lost at a ridge, though Schmidt and Gaudibert have traced it still farther in the same direction. It is at least 300 miles in length, and varies much in width and character, consisting in places of distinct crater-rows.

CRUGER.--A regular ring-plain E. of Fontana, 30 miles in diameter, with a dark floor, without detail, and comparatively low bright walls. There is a smaller but very conspicuous ring-plain (Cruger a) on the W. of it, to which runs a branch of the great Sirsalis cleft.

EICHSTADT.--A ring-plain, 32 miles in diameter, near the E. limb, S. of Rocca. It is the largest and most southerly of three nearly circular enclosures, without central mountains or any other details of interest. On the W. lies a great walled-plain with a very irregular border, containing several ring-plains and craters, and a crater-rill. Schmidt has named this formation DARWIN.

BYRGIUS.--A very irregular enclosure, about 40 miles in diameter, between Cavendish and the E. limb, with a lofty and discontinuous border, rising at one point on the E. to a height of 7000 feet above the floor. There are wide openings both in the N. and S. wall, and some ridges within. The border is broken on the E. by a crater, and on the W. by the well-known crater Byrgius A, from which a number of bright streaks radiate, mostly towards the E. One on the W. extends to Cavendish, and another to Mersenius, traversing the ring-plain Cavendish C. North-east of Byrgius there is a mountain arm which includes a peak 13,000 feet in height.

PIAZZI.--A walled-plain, about 90 miles in length, some distance S.E. of Vieta, with a complex broken border, including several depressions on the N.W., rising to about 7000 feet above a rather dark interior, on which there is a prominent central mountain.

LAGRANGE.--A larger but similar formation, 100 miles in diameter, associated with the last on the N.E., with a complex terraced border, including peaks of 9000 feet, a bright crater on the W., and a ring-plain on the N.W. The inner slope of the E. wall is a fine object at sunrise, when libration is favourable. The floor is dark and devoid of detail.

BOUVARD.--A great irregular enclosure, which appears to be still larger than Lagrange, S.E. of Piazzi, and close to the limb. It is bounded by a very lofty rampart, rising at a peak on the W. to 10,000 feet. It has a fine central mountain.

INGHIRAMI.--A very remarkable ring-plain, 60 miles in diameter, E. of Schickard, with a bright, broad, and nearly continuous border, terraced within, and intersected on the N.E. by narrow valleys, one of which is prolonged over the floor and extends to the central mountain. There are two curious dark spots on the N. side of the interior. Beyond the foot of the glacis on the S. a distinct cleft runs from a dusky spot to a group of small craters E. of Wargentin. There is a fine regular ring-plain with a small central mount W. of Inghirami.

PINGRE.--A ring-plain, about 18 miles in diameter, between Phocylides and the limb.

HAUSEN.--A ring-plain, close to the limb, N. of Bailly, which, but for its position, would be a fine object. It is, however, never sufficiently well placed for observation.

BAILLY.--One of the largest wall-surrounded plains on the moon, almost a "sea" in miniature, extending 150 miles from N. to S., and fully as much from W. to E. When caught at a favourable phase, it is, despite its position, especially worthy of scrutiny. The rampart on the W., of the linear type, is broken by several bright craters. On the S.W. two considerable overlapping ring-plains interfere with its continuity. On the S.E. several very remarkable parallel curved valleys traverse the border. The E. wall, which at one point attains a height of nearly 15,000 feet, is beautifully terraced. The floor on the eastern side includes several ring-plains (some of which are of a very abnormal type), many ridges, and two delicate dark lines, crossing each other near the S. end, probably representing clefts.

LEGENTIL.--A large walled-plain, close to the limb, S. of Bailly.

FOURTH QUADRANT

WEST LONGITUDE 90 deg. TO 60 deg.

KASTNER.--A large walled-plain at the S. end of the Mare Smythii, too near the limb for satisfactory observation.

MACLAURIN.--The principal member of a group of irregular ring-plains on the W. side of the Mare Foecunditatis, a little S. of the lunar equator. Schmidt shows no details within it, except a small crater on the E. side of the floor.

WEBB.--A ring-plain E. of Maclaurin, about 14 miles in diameter, with a dusky floor, enclosed by a bright rim, on the N.E. side of which there is a small crater. Schmidt seems to have overlooked the central hill.

LANGRENUS.--This noble circumvallation, the most northerly of the meridional chain of immense walled-plains, extending for more than 600 miles from near the equator to S. lat. 40 deg., would, but for its propinquity to the limb, rank with Copernicus (which in many respects it resembles) among the most striking objects on the surface of the moon. Its length is about 90 miles from N. to S., and its breadth fully as much. In shape it approximates very closely to that of a foreshortened regular hexagon. The walls, which at one point on the E. rise to an altitude of nearly 10,000 feet, are continuous, except on this side, where they are broken by the interference of an irregular depression, and on the extreme S., where they are intersected by cross-valleys. Within, the terraces are remarkably distinct, and the intervening valleys strongly marked. The brilliant compound central mountain rises at its loftiest peak to a height of more than 3000 feet. On the N. of it is an obscure circular ring, which may possibly merely represent a fortuitous combination of ridges, though it has all the appearance of a modified ring-plain. On the Mare, some distance N.E. of the formation, is a group of three ring-plains, with two small craters (associated with a ridge) on the N. of them. Two of the more westerly of these objects have prominent central mountains, and the third a very dark interior. At least three bright streaks originate on the E. flank of Langrenus, which, diverging widely, traverse the Mare Foecunditatis.

[FLATTENINGS ON THE MOON'S WESTERN LIMB.--About thirty years ago, the Rev. Henry Cooper Key drew attention to certain flattenings which he had noted on the W. limb, which are very apparent under favourable conditions of libration. Their position cannot be closely defined, but the principal deviation from circularity extends from about S. lat. 10 deg. to the region on the limb opposite the S. border of the Mare Crisium.]

VENDELINUS.--The second great enclosure pertaining to the meridional chain--a magnificent walled-plain of about the same dimensions as the last. It is bounded by a very irregular rampart, which, under evening illumination, is especially noteworthy, though nowhere approaching the altitude of that of Langrenus. Its continuity on the W. is broken by the great ring-plain Vendelinus C, about 50 miles in diameter, a formation resembling Langrenus in miniature. This is hexagonal in shape, and has many rings and depressions on its W. wall. South of Vendelinus C, the wall of Vendelinus runs up in a bold curve to the fine terraced ring- plain Vendelinus B, and is surmounted by a bright serpentine crest, and traversed by several valleys running down the slope to the floor. B has a small crater on its N. wall, and another in the interior. There is a wide gap in the S. border of Vendelinus, which is partially occupied by another somewhat smaller ring-plain, bounded by a southerly extension of the E. wall, which includes on its outer slope many craters and other depressions, and abuts near its N. end on the large ring-plain Vendelinus A, which has a prominently terraced wall and a large bright central mountain. Between A and C extends a plateau that may be regarded as the N. limit of the formation, including, among other minor details, a fine cleft, which traverses it from N. to S., and ultimately extends to a group of craters on the floor. On the S. side of the interior is one large ring-plain, flanked on the W. by two small craters. Near the N. end are many bright little craters, many of them unrecorded. Vendelinus C is bordered on the E. by two large semicircular formations with low walls extending on to the floor. Mr. W.H. Maw and others have detected many minute depressions in connection with these curious objects; and N. of them, on the outer slope of C, where it runs out to the level of the plateau, I have seen the surface at sunset riddled like a sieve with craterlets and little pits. There is an irregular ring-plain N. of A, with linear walls, and another, much smaller and brighter, on the N. of this, standing a little beyond the N. limits of Langrenus.

LA PEYROUSE.--A much foreshortened walled-plain, 41 miles in diameter, close to the limb, S.W. of Langrenus. There is a longitudinal ridge on the floor. Between it and Langrenus are two large ring-plains with central mountains, and on the N.E., La Peyrouse A, a bright crater, adjoining which is La Peyrouse DELTA, one of the most brilliant spots on the moon.

ANSGARIUS.--A ring-plain, 50 miles in diameter, still nearer to the limb than the last.

BEHAIM.--A great ring-plain, 65 miles in diameter, S. of Ansgarius, and connected with it by ridges. It has lofty walls and a central mountain.

HECATAEUS.--An immense walled-plain, 115 miles in length, on the S.W. of Vendelinus, with a very irregular rampart and a conspicuous central mountain. It is flanked E. and W. by other large enclosures, which can only be seen to advantage when libration is favourable.

W. HUMBOLDT.--Though close to the limb, this enormous wall-surrounded plain, some 130 miles in extreme length, and estimated to have an area of 12,000 square miles, is well worth observing under suitable conditions. It ranks among the largest formations of its class, and in many respects resembles Bailly on the S.E. limb. At one point on the E. a peak rises to 16,000 feet, and on the opposite side there are peaks nearly as high. The floor contains some detail--a crater, nearly central, associated with ridges, and two dark spots, one at the S. and the other at the N. end.

PHILLIPS.--Abuts on the E. side of W. Humboldt. It is a walled-plain, about 80 miles in length, with a border much broken on the E., and terraced within on the opposite side. There are many hills and ridges on the floor.

LEGENDRE.--A fine ring-plain, 46 miles in diameter, on the S.E. of the last. According to Schmidt, there is a crater on the S. side of the floor. There is a small ring-plain, ADAMS, on the S.

PETAVIUS.--The third member of the great meridional chain: a noble walled-plain, with a complex rampart, extending nearly 100 miles from N. to S., which encloses a very rugged convex floor, traversed by many shallow valleys, and includes a massive central mountain and one of the most remarkable clefts on the visible surface. To observe these features to the best advantage, the formation should be viewed when its W. wall is on the evening terminator. At this phase a considerable portion of the interior on the N. is obscured by the shadow of the rampart, but the principal features on the S. half of the floor, and on the broad gently- shelving slope of the W. wall, are seen better than under any other conditions. The border is loftiest on the E., where the ring-plain Wrottesley abuts on it. It rises at this point to nearly 11,000 feet, while on the opposite side it nowhere greatly exceeds 6000 feet above the interior. The terraces, however, on the W. are much more numerous, and, with the associated valleys, render this section of the wall one of the most striking objects of its class. The N. border is conspicuously broken by the many valleys from the region S. of Vendelinus, which run up to and traverse it. On the S., also, it is intersected by gaps, and in one place interrupted by a large crater. There is a remarkable bifurcation of the border S. of Wrottesley. A lower section separates from the main rampart and, extending to a considerable distance S.E. of it, encloses a wide and comparatively level area which is crossed by two short clefts. The central mountains of Petavius, rising at one peak to a height of nearly 6000 feet above the floor, form a noble group, exceeding in height those in Gassendi by more than 2000 feet. The convexity of the interior is such that the centre of it is about 800 feet higher than the margin, under the walls; a protuberance which would, nevertheless, be scarcely remarked _in situ_, as it represents no steeper gradient than about 1 in 300 on any portion of its superficies. The great cleft, extending from the central mountains to the S.E. wall, and perhaps beyond, was discovered by Schroter on September 16, 1788, and can be seen in a 2 inch achromatic. In larger instruments it is found to be in places bordered by raised banks.

WROTTESLEY.--A formation, about 25 miles in diameter, closely associated with the E. wall of Petavius, the shape of which it has clearly modified. Its border on the E., of the linear type, rises nearly 9000 feet above a light interior, where there is a small bright central mountain and some mounds. There is a prominent valley running along the inner slope of the W. wall.

PALITZSCH.--If this extraordinary formation is observed when the moon is about three days old, it resembles a great trough, or deep elongated gorge flanking the W. wall of Petavius, though it is a true ring-plain, albeit of a very abnormal type, about 60 miles in length and 20 miles in breadth, with a somewhat dusky interior. On the outer slope of its W. wall is a bright ring-plain with a lofty border and a central mountain.

HASE.--An irregular formation, about 50 miles in diameter, on the S.W. of Petavius, with which it is connected by extensions of the W. and E. walls of the latter. Its rampart, some 7000 feet above the floor, is broken by depressions on the W.; and on the S. is bounded by a smaller ring-plain with still loftier walls. Schmidt shows a large crater and three smaller ones on the W. side of the floor.

MARINUS.--A ring-plain on the N.E. side of the Mare Australe, between Furnerius and the limb.

FURNERIUS.--The fourth and most southerly component of the great meridional chain of walled-plains, commencing on the N. with Langrenus: a fine but irregular enclosure, about 80 miles in extreme length and much more in breadth. Its rampart is very lofty, and tolerably continuous on the N. and W., but on the other sides is interrupted by small craters and depressions. At peaks on the E. it attains a height of more than 11,000 feet above the interior, and there are other peaks rising nearly as high. There is a ring-plain (Furnerius B) with a central hill, on the E. side of the floor, and numerous craters and crater-pits in other parts of it. On the N.W. side of B there is a short cleft, on the W., a well-marked crater-row, and on the E. a long rill-valley. The very brilliant crater (Furnerius A) on the N.E. glacis is the origin of two fine light streaks, one extending S. for more than 100 miles, and the other in the opposite direction for a great distance.

FRAUNHOFER.--A ring-plain, S. of Furnerius, about 30 miles in diameter, with a regular border rising about 5000 feet above the floor. A smaller ring-plain abuts on the N.E. side of it, which has slightly disturbed its wall.

OKEN.--A large enclosure in S. lat. 43 deg. with broken irregular walls. It is too near the limb for observation.

VEGA.--Schmidt represents this peculiar formation, situated S.E. of Oken, as having a regular curved unbroken rampart on the E., while the opposite border is occupied by four large partially overlapping ring-plains, two of which contain small craters. The floor is devoid of detail.

PONTECOULANT.--A great irregular walled plain, about 100 miles in length, near the S.W. limb, with a border rising in places to a height of 6000 feet above the floor.

HANNO.--A smaller and more regular enclosure, adjoining Pontecoulant on the N.W., and still nearer the limb.

WEST LONGITUDE 60 deg. TO 40 deg.

MESSIER.--The more westerly of a remarkable pair of bright craters, about 9 miles in diameter, standing in an isolated position in the Mare Foecunditatis just S. of the Equator. Madler represents them as similar in every respect, but Webb, observing them in 1855 and 1856 with a 3 7/10 achromatic, found them very distinctly different,--Messier, the more westerly, being not only clearly smaller than its companion, but longer from W. to E. than from N. to S., as it undoubtedly is at the present time. Messier A, however, as the companion is termed, though larger, is certainly not circular, as sometimes shown, but triangular with curved sides. It is just possible that change may have occurred here, for Madler carefully observed these objects more than three hundred times, and, it may be presumed, under very different phases. Messier A is the origin of two slightly divergent light streaks, resembling a comet's tail, which extend over the Mare towards its E. border N. of Lubbock, and are crossed obliquely by a narrower streak. Messier and Messier A stand near the S. and narrowest end of a tapering curved light area. There is a number of craterlets and minute pits in the neighbourhood, and under a high light two round dusky spots are traceable in connection with the "comet" marking, one just beyond its northern, and the other beyond its southern border, near its E. extremity.

LUBBOCK.--A brilliant little crater, about 4 or 5 miles in diameter, near the E. coast-line of the Mare Foecunditatis. The region E. of this object is particularly well worthy of scrutiny under a low sun, on account of the variety of detail it includes. On the S.E. run three fine parallel clefts, originating near the N. end of the Pyrenees.

GUTTEMBERG.--A very fine ring-plain of peculiar shape, about 45 miles in width, with a lofty wall, broken on the N.W. by another ring-plain some 14 miles in diameter, and on the S.E. by a small but distinct crater. The border presents a wide opening towards the S., which is traversed by a number of longitudinal valleys, both the E. and W. sections of the wall being prolonged in this direction. A fine crater-row runs round the outer slope of the E. wall, from the crater just mentioned to the N. side of the formation. It is best seen when the W. wall is on the evening terminator. There is also a broad valley on the S. prolongation of the W. wall. The central mountain is bright but not large. A cleft crosses the N.W. side of the floor. North of Guttemberg there is a curious oblong formation with low walls, connected with the N.E. border by a ridge, and with the N. border by a remarkable row of depressions, situated on a mound; and beyond this object on the E. are three parallel clefts running towards the N.E. On the W. will be found some of the clefts belonging to the Goclenius rill-system. In the rugged region S.E. of the formation is a peculiar low ring with a very uneven floor and a large central hill. The E. wall of Guttemberg may be regarded as forming a portion of the Pyrenees Mountains.

GOCLENIUS.--A ring-plain, about 28 miles in diameter, bearing much resemblance to Plinius in form and size, and, like this formation, associated with a fine system of clefts. The lofty rampart, tolerably continuous on the W., is broken on the S.W. by a bright crater, and on the N.W. by a remarkable triangular depression. It is also traversed by a delicate valley extending from the crater on the S.W. to another on the N.W. border; and at a point a little W. of the first crater is dislocated by an intrusive mass of rock. There are several gaps on the E. and many spurs and irregularities in outline both within and without. A great portion of the N. wall is linear, and joins the E. section nearly at right angles. West of the triangular depression it appears to be partially wrecked, indications of the destruction being very evident if it be observed when the E. wall is near the morning terminator. The small bright central mountain is remarkable for its curious oblong shadow. Two clefts traverse the interior of Goclenius. (1) Originates at the S. wall, E. of the crater, and runs E. of the central mountain to the N. wall; (2) crosses the debris of the ruined N.W. border, runs parallel to the first, and extends nearly to the centre of the floor, (1) Re-appears at the foot of a mound outside the N. wall, and, after crossing the outer W. slope of the great ring-plain on the N.W. wall of Guttemberg, runs to the W. side of an oblong formation N. of it. There are two other clefts, closely parallel and W. of this, traversing the Mare, and terminating among the mountains on the N.W. These are crossed at right angles by what appears to be a "fault," running in a N.W. direction from the W. side of Guttemberg.

MACCLURE.--One of a curious group of formations situated in the Mare Foecunditatis some distance S.W. of Goclenius. It is a bright ring-plain, about

15 miles in diameter, with a narrow gap in the N.E. wall and a small central hill. A prominent ridge runs up to the N. border; and on the S.W. a rill-valley may be traced, extending S. to a bright deep little crater W. of Cook.

CROZIER.--A conspicuous ring-plain a few miles N.N.W. of MacClure, and of about the same size. It has a faint central hill. Neison refers to two long straight streaks extending from Crozier towards Messier.

BELLOT.--A brilliant little ring-plain N.E. of Crozier.

COOK.--A ring-plain, about 25 miles in diameter, on the E. side of the Mare Foecunditatis in S. lat. 17 deg., with low and (except on the S.E.) very narrow walls. There is a small circular depression on the S. border, and a prominent crater on the W. side of the dark interior. On the S.S.E. is the curiously shaped enclosure Cook d, with very bright broad lofty walls and a fine central mountain. On the plain W. of Cook is a conspicuous crater-row, consisting of six or seven craters, diminishing in size in both directions from the centre.

COLOMBO.--A fine ring-plain, about 50 miles in diameter, situated in the highlands separating the Mare Foecunditatis and the Mare Nectaris. The wall, rising at one place to a height of 8000 feet above the floor, is very complicated and irregular, being traversed within by many terraces, and almost everywhere by cross-valleys. Its shape is greatly distorted by the large ring-plain a, which abuts on its N.E. flank. It loses its individuality altogether on the S., its place being occupied by two large depressions, and lofty mountains trending towards the S.E. In the centre there are several distinct bright elevations.

MAGELHAENS.--The more northerly and the larger of a pair of ring-plains between Colombo and Goclenius, with a bright and somewhat irregular though continuous border. The dark interior includes a small central mountain. Its companion on the S.W., Magelhaens a, slightly overlaps it. This also has a central hill, and a crater on the outer slope of its E. wall.

SANTBECH.--A very prominent ring-plain, 46 miles in diameter, on the S.E. side of the Mare Foecunditatis, W. of Fracastorius. The continuity of its fine lofty rampart is broken on the W., where it rises nearly 10,000 feet above the floor, by a brilliant little crater just below the crest, and by a narrow gap on the S. The wall on the E. towers to a height of 15,000 feet above the interior. On its broad outer slope, near the summit, there is a fine crater, and S. of this running obliquely down the slope a distinct valley. On the N.E., where the glacis runs down to the level of the surrounding plain, there is a large crateriform object with a broken N. border, and a small crater opposite the opening. A long coarse valley runs from this latter object in a N.E. direction to the region W. of Bohnenberger. Santbech contains a prominent central peak.

BIOT.--A brilliant little ring-plain, scarcely more than 7 miles in diameter, standing in an isolated position in the Mare Foecunditatis N.E. of Wrottesley. There is a number of bright streaks in its neighbourhood; and a few miles E. of it, in the hilly region W. of Santbech, another conspicuous crater of about the same size.

BORDA.--A ring-plain about 25 miles in diameter, S.S.W. of Santbech, with a rampart low on the N. and S., but elsewhere of considerable height, and a very conspicuous central mountain. A wide deep valley flanked by lofty mountains extends from the N. wall for many miles towards the N.W. It is an especially noteworthy object when the W. wall of Santbech is on the evening terminator, as its somewhat winding course, indicated by the bright summit-ridges of the bordering mountains, can be followed some hours before either the interior of the valley or the region between it and Santbech are in sunlight. Among the mountains W. of Borda there is a peak more than 11,000 feet in height.

SNELLIUS.--A very fine ring-plain, 50 miles in diameter, S.E. of Petavius, with terraced walls, considerably broken on the S.E. by craters, &c. It rises on the E. nearly 7000 feet above a dark floor, which contains a central mountain. N.E. of Snellius is a smaller ring- plain (Snellius a), and due E. a curious rough plateau, bordered on the N. and S. by a number of small craters.

STEVINUS.--A somewhat larger ring-plain, S. of Snellius, with a border rising on the S. to more than 11,000 feet above a dark interior, which includes a bright central mountain.

REICHENBACH.--A very abnormally-shaped ring-plain, about 30 miles in diameter, with a rampart nearly 12,000 feet high. The border is broken on the W., S., and E. by craters and depressions, and on the N. is flanked by two overlapping ring-plains, a and b. On the S.W. lies a magnificent serpentine valley, fully 100 miles in length and about 12 miles in breadth at the N. end, but gradually diminishing as it runs southwards, till it reaches a depression N. of Rheita, where it terminates: here is scarcely more than 4 miles wide.

RHEITA.--A formation, about 35 miles in diameter, S. of Reichenbach, with regular lofty walls, rising at a peak on the N.E. to a height of more than 14,000 feet above the interior, on which there is a small but prominent central mountain, a smaller elevation W. of the centre, and two adjoining craters at the foot of the S. wall. On the E. originates another fine valley, very similar to that already mentioned in connection with Reichenbach. It runs in a S.S.W. direction, is about 100 miles in length, and, in its widest part, is about 12 miles across. Like the Reichenbach valley, it terminates at a small crater-like object, which has a border broken down on the side facing the valley, and a small central hill. About midway between its extremities, this great gorge is crossed by a wall of rock, like a narrow bridge.

JANSSEN.--An immense irregular enclosure, reminding one of the very similar area, bordered by Walter, Lexell, Hell, &c., in the third quadrant. It extends about 150 miles from E. to W., and more than 100 from N. to S., its limits on the N. being rather indefinite. Its very rugged humpy surface includes one great central mountain, and innumerable minor hills and ridges, craters, and crater-pits; but the principal feature is the magnificent curved rill-valley running from the S. side of Fabricius across the rough expanse to the S. side. This fine object, very coarse on the N., passes the central mountain on the E. side, and becomes gradually narrower as it approaches the border; before reaching which, another finer cleft branches from it on the W., and also runs to the S. side of the plain.

LOCKYER.--A prominent deep ring-plain, 32 miles in diameter, with massive bright lofty walls, standing just outside the S.E. border of Janssen. Schmidt shows a minute crater on the S. rim. I have seen a crater within, at the inner foot of the W. wall, and a central peak.

FABRICIUS.--A ring-plain, 55 miles in diameter, with a lofty terraced border, rising on the S.W. to a height of nearly 10,000 feet above the interior. It is partially included by the rampart of Janssen, and the great rill-valley on the floor of the latter appears to cut through its S. wall. There is a long central mountain on the floor, with a prominent ridge extending along the E. side of it. W. of Fabricius (between it and the border of Janssen) lies a very irregular enclosure, with three distinct craters within it; and on the E., running from the wall to the E. side of Janssen, is a straight narrow valley. Both Fabricius and Janssen should be viewed under a low morning sun.

STEINHEIL.--A double ring-plain, W. of Janssen, 27 miles in diameter. The more easterly formation sinks to a depth of nearly 12,000 feet below the summit of the border.

METIUS.--This ring-plain, of about the same size as Fabricius, but with a still loftier barrier, abuts on the N. wall of this formation, and has caused a very obvious deformation in its contour. It is prominently terraced internally, and on the W. the wall rises at one peak to a height of 13,000 feet above the floor, which contains a deep crater on the W. of the centre, and many ridges.

BIELA.--A considerable ring-plain, about 55 miles in diameter, S.W. of Janssen, with a wall broken on the N.W., S., and E. by rings and large enclosures. There is a central mountain, but apparently no other details on the floor.

ROSENBERGER.--This formation, about 50 miles in diameter, is one of the remarkable group of large rings to which Vlacq, Hommel, Pitiscus, &c., belong. Its walls, though of only moderate altitude, are distinctly terraced. In addition to a prominent central mountain (E. of which Schmidt shows two craters), there is a large crater on the S. side of the floor, and many smaller craters and crater-pits.

HAGECIUS.--The most westerly member of the Vlacq group of formations. It is situated on the S.W. of Rosenberger, and is about 50 miles in diameter. The rampart on the E. is continuous and of the normal type, but on the opposite side is broken by a number of smaller rings.

WEST LONGITUDE 40 deg. TO 20 deg.

CENSORINUS.--A brilliant little crater, with very bright surroundings, in the Mare Tranquilitatis, nearly on the moon's equator, in W. long. 32 deg. 22 min. Another smaller but less conspicuous crater adjoins it on the W. On the Mare to the S. extends a delicate cleft which trends towards the Sabine and Ritter rill system.

CAPELLA.--Forms with Isodorus, its companion on the E. (which it partially overlaps), a very noteworthy object. It is about 30 miles in diameter, with finely terraced walls, broken on the S.W. by broad intrusive rill-valleys. The rampart on the N.E. is also cut through by a magnificent valley, which extends for many miles beyond the limits of the formation. There is a fine central mountain, on which M. Gaudibert discovered a crater, the existence of which has been subsequently verified by Professor Weinek on a Lick observatory negative.

ISODORUS.--The rampart of this fine ring-plain, which is of about the same size as Capella, rises at a peak on the W. to a height of more than 13,000 feet above the interior, which, except a small bright crater at the foot of the E. wall and a smaller one adjoining it on the N., contains no detail. The region between Isodorus and the equator includes many interesting objects, among them Isodorus b, an irregular formation open towards the N., and containing several craters.

BOHNENBERGER.--A ring-plain about 22 miles in diameter, situated on the W. side of the Mare Nectaris, under the precipitous flanks of the Pyrenees, whose prominent shadows partially conceal it for many hours after sunrise. The circular border is comparatively low, and, except on the N., continuous. Here there is a gap, and on the W. of it an intrusive mass of rock. From its very peculiar shadow at sunrise, the wall on the E. appears to be very irregular. The club-shaped central mountain is of considerable size, but not conspicuous. S. of Bohnenberger stands the very attenuated ring, Bohnenberger A. It is of about the same diameter, has a large deep crater on its N. rim, and a smaller one, distinguished with difficulty, on its S.E. rim. On the N. of Bohnenberger there is a bright little ring-plain connected with the formation by a lofty ridge, under the E. flank of which Schmidt shows a crater-chain. An especially fine cleft originates on the E. side of this crater, which, following an undulating course over the Mare Nectaris, terminates at Rosse, N. of Fracastorius.

TORRICELLI.--A remarkable little formation in the Mare Tranquilitatis, N. of Theophilus, consisting of two unequal contiguous craters ranging from W. to

E., whose partition wall has nearly disappeared, so that, under a low sun, when the interior of both is filled with shadow, the pair resemble the head of a javelin. The larger, western, ring is about 10 miles in diameter, and the other about half this size. There is a gap in the W. wall of the first, and a long spur projecting from its S. side; and a minute crater on the S. border of the smaller object. Torricelli is partially enclosed on the S. by a circular arrangement of ridges. There is a delicate cleft running in a meridional direction on the Mare, E. of the formation, and another on the N., running from W. to E.

HYPATIA.--A ring-plain, about 30 miles in extreme length, of very abnormal shape, on the E. side of the Mare, N.N.E. of Theophilus, with a wall rising at a peak on the E. to a height of more than 7000 feet above a dusky floor, which does not apparently contain any detail. A small crater breaks the uniformity of the border on the W. Beyond the wall on the S.E. lies the fine bright crater Hypatia A, with another less prominent adjoining it on the S.W.

THEOPHILUS.--The most northerly of three of the noblest ring-mountains on the visible surface of the moon, situated on the N.E. side of the Mare Nectaris. It is nearly 64 miles in diameter, and is enclosed by a mighty rampart towering above the floor at one peak on the W. to the height of 18,000 feet, and at two other peaks on the opposite side to nearly 16,000 and 14,000. The border, though appearing nearly circular with low powers, is seen, under greater magnification, to be made up of several more or less linear sections, which give it a polygonal outline. It is prominently terraced within, the loftier terraces on the W. rising nearly to the height of the crest of the wall, and including several craters and elongated depressions. On the W. glacis is a row of large inosculating craters; and near its foot, S.E. of Madler, a short unrecorded rill- valley. The magnificent bright central mountain is composed of many distinct masses surmounted by lofty peaks, one of which is about 6000 feet above the floor, and covers an area of at least 300 square miles. Except a distinct crater on the S.W. quarter, this appears to be the only object within the ring.

CYRILLUS.--The massive border of Theophilus partially overlaps the N.W. side of this great walled-plain, which is even more complex than that of its neighbour, and far more irregular in form, exhibiting many linear sections. Its crest on the S.E. is clearly inflected towards the interior, a peculiarity that has already been noticed in connection with Copernicus and some other objects. On the inner slope of this wall there is a large bright crater, in connection with which have been detected two delicate rills extending to the summit. I have not seen these, but one of the crater-rows shown by Schmidt, between this crater and the crest, has often been noted. The N.E. wall is very remarkable. It appears to be partially wrecked. If observed at an early stage of sunrise, a great number of undulating ridges and rows of hillocks will be seen crossing the region E. of Theophilus. They resemble a consolidated stream of "ropy" lava which has flowed through and over the wall and down the glacis. The arrangement of the ridges within Cyrillus is very noteworthy, as is also the triple mountain near the centre of the floor. The fine curved cleft thereon traverses the W. side, sweeping round the central mountains, and then turning to the south. I have only occasionally seen it in its entirety. There are also two oblong dark patches on the S. side of the interior. The S. wall of Cyrillus is broken by a narrow pass opening out into a valley situated on the plateau which bounds the W. side of the oblong formation lying between it and Catherina, and overlooking a curious shallow square-shaped enclosure abutting on the S.W. side of Cyrillus.

CATHERINA.--The largest of the three great formations: a ring-plain with a very irregular outline, extending more than 70 miles in a meridional direction, and of still greater width. The wall is comparatively narrow and low on the N.E. (8000 feet above the floor), but on the N.W. it rises to more than double this height, and is broken by some large depressions. The inner slope on the S.E. is very gentle, and includes two bright craters, but exhibits only slight indications of terraces. The most remarkable features on an otherwise even interior are the large low narrow ring (with a crater within it), occupying fully a third of the area of the floor, and a large ring-plain on the S. side.

MADLER.--The interest attaching to this formation is not to be measured by its size, for it is only about 20 miles in diameter, but by the remarkable character of its surroundings. Its bright regular wall, rising 6000 feet on the E. and only about half as much on the W., above a rather dark interior, is everywhere continuous, except at one place on the N. Here there is a narrow gap (flanked on the E. by a somewhat obscure little crater) through which a curious bent ridge coming up from the N. passes, and, extending on to the floor, expands into something resembling a central mountain. Under a high sun Madler has a very peculiar appearance. The lofty E. wall is barely perceptible, while the much lower W. border is conspicuously brilliant; and the E. half of the floor is dark, while the remainder, with two objects representing the loftier portions of the intrusive ridge, is prominently white. Under an evening sun, with the terminator lying some distance to the W., a very remarkable obscure ring with a low border, a valley running round it on the W. side, and two large central mounds, may be easily traced. This object is connected with Madler by what appears to be under a higher sun a bright elbow-shaped marking, in connection with which I have often suspected a delicate cleft. Between the obtuse-angled bend of this object and the W. wall of Madler, two large circular dark spots may be seen under a high sun; and on the surface of the Mare N. of it, a great number of delicate white spots.

BEAUMONT.--A ring-plain about 30 miles in diameter, on the S.E. side of the Mare Nectaris, midway between Theophilus and Fracastorius, with the N.E. side of which it is connected by a chain of large depressions. Its border is lofty, regular, and continuous on the S. and E., but on the W. it is low, and on the N. sinks to such a very inconsiderable height that it is often scarcely traceable. It exhibits two breaks on the S.W., through one of which passes a coarse valley that ultimately runs on the E. side of the depressions just referred to. The interior is pitted with many craters, one on the W. side being shallow but of considerable size. I once counted twenty with a 4 inch Cooke achromatic, and Dr. Sheldon of Macclesfield subsequently noted many more. A ridge, prominent under oblique light, follows a winding course from the N.W. side of Beaumont to the W. side of Theophilus, and there is another lower ridge E. of it. Between them is included a region covered with minute hillocks and asperities. Among these objects are certain dusky little crater-cones, which Dr. Klein of Cologne regards as true analogues of some terrestrial volcanoes. They are very similar in character to those, already alluded to, in the dusky area between Copernicus and Gambart.

KANT.--A conspicuous ring-plain, 23 miles in diameter, situated in a mountainous district E. of Theophilus, with lofty terraced walls and a bright central peak. Adjoining it on the W. is a mountain mass, projecting from the coast-line of the Mare, on which there is a peak rising to more than 14,000 feet above the surface.

FRACASTORIUS.--This great bay or inflexion at the extreme S. end of the Mare Nectaris, about 60 miles in diameter, is one of the largest and most suggestive examples of a partially destroyed formation to be found on the visible surface. The W. section of the rampart is practically complete and unbroken, rising at one peak to a height of 6000 feet above the interior. It is very broad at its S. end, and its inner slope descends with a gentle gradient to the floor. Towards the N., however, it rapidly decreases in width, but apparently not in altitude, till near its bright pointed N. extremity. Under a low sun, some long deformed crateriform depressions may be seen on the slope, and a bright little crater on the crest of the border near its N. end. The southern rampart is broken by three large craters, and a fine valley, running some distance in a S. direction, which diminishes gradually in width till it ultimately resembles a cleft, and terminates at a small crater. The E. border is very lofty and irregular, rising at the N. corner of the large triangular formation, which is such a prominent feature upon it, to a height of 7000 feet, and at a point on the S.E. to considerably more than 8000 feet above the floor. N. of the former peak it becomes much lower and narrower, and is finally only represented by a very attenuated strip of wall, hardly more prominent than the brighter portions of the border of Stadius at sunrise, terminating at an obscure semi-ring-plain. Between this and the pointed N. termination of the W. border there is a wide gap, open to the north for a space of about 30 miles, appearing, except under very oblique illumination, as smooth and as devoid of detail as the grey surface of the Mare Nectaris itself. If, however, this interval is observed at sunrise or sunset, it is seen to be not quite so structureless as it appears under different conditions, for a number of mounds and large humpy swellings, with low hills and craterlets, extend across it, and occupy a position which we are justified in regarding as the site of a section of the rampart, which, from some cause or other, has been completely destroyed and overlaid with the material, whatever this may be, of the Mare Nectaris. The floor of Fracastorius is, as regards the light streaks and other features upon it, only second in interest to those of Plato and Archimedes, and will repay systematic observation. Between thirty and forty light spots and craters have been recorded on its surface, most of them, as in these formations, being situated either on or at the edges of the light streaks. On the higher portion of the interior (near the centre) is a curious object consisting apparently of four light spots, arranged in a square, with a craterlet in the middle, all of which undergo (as I have pointed out elsewhere) notable changes of aspect under different phases. There are at least two distinct clefts on the floor, one running from the W. wall towards the centre, and another on the S.E. side of the interior. The last throws out two branches towards the S.W.

ROSSE.--A fine bright deep crater in the Mare Nectaris, N. of the pointed termination of the W. wall of Fracastorius, with which it is connected by a bold curved ridge, with a crater upon it. A ray from Tycho, striking along the E. wall of Fracastorius passes near this object. A rill from near Bohnenberger terminates at this crater.

POLYBIUS.--A ring-plain, about 17 miles in diameter, in the hilly region S.E. of Fracastorius. The border is unbroken, except on the N., where it is interrupted by a group of depressions. There is a long valley on the S.W., at the bottom of which Schmidt shows a crater-chain.

NEANDER.--This ring-plain, 34 miles in diameter, a short distance W.S.W. of Piccolomini, has a somewhat deformed rampart, which, however, except on the N., where there is a narrow gap occupied by a small crater, is continuous. It rises on the E. nearly 8000 feet above the floor, on which there is a central mountain about 2500 feet high. Schmidt shows some minor hills, a large crater on the N.E. side, and three smaller craters in the interior.

PICCOLOMINI.--A ring-plain of a very massive type, about 57 miles in diameter, S. of Fracastorius, with complex and prominently terraced walls, surmounted by very many peaks; one of which on the E. attains a height of 14,000 feet, and another, N. of it, on the same side, an altitude of 15,000 feet above the interior. The crest of this grand rampart is tolerably continuous, except on the S.W., where, for a distance of twenty miles or more, its character as regards form and brightness is entirely changed. Under a low sun, instead of a continuous bright border, we note a wide gap occupied by a dusky rugged plateau, which falls with a gentle gradient to the floor, and is traversed by three or four parallel shallow valleys running towards the S. I can recall no lunar formation which presents an appearance at all like this: one is impressed with the idea that it has resulted from the collapse of the upper portion of the wall, and the flow of some viscous material over the wreck and down the inner slope. The difference between the reflective power of this matter, whatever may be its nature, and the broad bright declivities of the inner slopes, are beautifully displayed at sunset. The cross-valleys are more easily traced under low morning illumination; but to appreciate the actual structure of the wall, it should be observed under both phases. The N.W.

section of the border includes many "pockets," or long elliptical depressions, which at an early stage of sunrise give a scalloped appearance to the crest. Except the great bright central mountain with its numerous peaks, there does not appear to be any prominent detail on the floor. There is a large ring-plain beyond the foot of the glacis on the W. with two craters on the E. side of it, another on the S., and a fine rill-valley running up to its N. side from near the crest of the W. wall. On the N. side of Piccolomini is a remarkable group of deformed and overlapping enclosures, mingled with numberless craters and little depressions. The plain on the N.E. is crossed by a fine cleft.

PONS.--A complete formation of irregular shape, about 20 miles in greatest diameter, on the S.E. side of the Altai range, in W. long. 21 deg. It consists of a crowd of rings and craters enclosed by a narrow wall.

STIBORIUS.--An elongated ring-plain, about 22 miles in diameter, S. of Piccolomini, with a lofty wall, broken in one place on the N. by a very conspicuous crater. Schmidt shows a distinct crater in the centre of the floor. I have only seen a central mountain in this position. There is a large crater on the N.W., a ring-plain on the S.W. side, and a multitude of little craters on the surrounding plain.

RICCIUS.--A ring-plain, 51 miles in diameter, of a very irregular type, S.E. of the last. It is enclosed by a complex wall (which is in places double), broken by large rings on the S. The very conspicuous little ring-plain Riccius A is situated on the N. of it, and other less prominent features. The interior includes a bright crater and some smaller objects of the same class.

ZAGUT.--The most easterly of a group of closely associated irregular walled-plains, of which Lindenau and Rabbi Levi are the other members, all evidently deformed and modified in shape by their proximity. It is about 45 miles in diameter, and is enclosed by a wall which on the S.W. attains a height of about 9500 feet, and is much broken on the N. by a number of depressions. A large ring-plain, some 20 miles in diameter, occupies a considerable portion of the W. side of the interior; E. of which, and nearly central, there is a large bright crater, but apparently no other conspicuous details. On the S.E. side of Zagut lies an elliptical ring-plain, about 28 miles in diameter, named by Schmidt CELSIUS. The border of this is open on the N., the gap being occupied by a large crater, whose S. wall is wanting, so that the interiors of both formations are in communication.

LINDENAU.--This formation, about 35 miles in diameter, is bounded on the W. by a regular unbroken wall nearly 8600 feet in height; but which on the E. and N.E. is far loftier and more complex, rising to about 12,000 feet above the floor, consisting of four or more distinct ramparts, separated by deep valleys, and extending towards Rabbi Levi. Neison points out that under a high light Lindenau appears to have a bright uniform single wall. There is a small central mountain and some minor inequalities in the interior.

RABBI LEVI.--A larger but less obvious formation than either of its neighbours, Zagut and Lindenau, abutting on the S. side of them. It is about 55 miles in diameter, and is enclosed by a border somewhat difficult to trace in its entirety, except under oblique light. There are some large craters within it, of which one on the N. side of the floor is especially prominent.

NICOLAI.--A tolerably regular ring-plain, 18 miles in diameter, S. of Riccius, with a border, rising more than 6000 feet above a level floor, on the N. side of which Schmidt shows a minute crater. The bright plain surrounding this formation abounds in small craters; and on the W. is a number of curious enclosures, many of them overlapping.

VLACQ.--A member of a magnificent group of closely associated formations situated on the greatly disturbed area between W. long. 30 deg. and 45 deg. and S. lat. 50 deg. and 60 deg. It is 57 miles in diameter, and is enclosed by terraced walls, rising on the W. about 8000 feet, and on the E. more than 10,000 feet above the floor. They are broken on the S. by a fine crater. In addition to a conspicuous central peak, there are several small craters, and low short ridges in the interior.

HOMMEL.--Adjoins Vlacq on the S. It is a somewhat larger and a far more irregular formation. On every side except the W., where the border is unbroken, and descends with a gentle slope to the dark interior; ring- plains and smaller depressions encroach on its outline, perhaps the most remarkable being Hommel a on the N., which has an especially brilliant wall, that includes a conspicuous central mountain, a large crater, and other details. The best phase for observing Hommel and its surroundings is when the W. wall is just within the evening terminator.

PITISCUS.--The most regular of the Vlacq group. It is situated on the N.E. of Hommel (a curious oblong-shaped enclosure, Hommel h, with a very attenuated E. wall, and a large crater on a floor, standing at a higher level than that of Pitiscus, intervening). It is 52 miles in diameter, and is surrounded by an apparently continuous rampart, except on the E., where there is a crater, and on the S.W., where it abuts on Hommel h. Here there is a wide gap crossed by what has every appearance of being a "fault," resembling that in Phocylides on a smaller scale. There is a fine crater on the N. side of the interior connected with the S. wall by a bright ridge. Just beyond the E. border there is a shallow ring-plain of a very extraordinary shape.

NEARCH.--A ring-plain, about 35 miles in diameter, on the S.W. of Hommel, forming part of the Vlacq group.

TANNERUS.--A ring-plain, about 19 miles in diameter, between Mutus and Bacon. It has a central mountain.

MUTUS.--A fine but foreshortened walled plain, 51 miles in diameter. There are two ring-plains of about equal size on the floor, one on the N., and the other on the S. side. The wall on the W. rises to nearly 14,000 feet above the interior.

MANZINUS.--A walled plain, nearly 62 miles in diameter, with a terraced rampart rising to a height of more than 14,500 feet above the interior. Schmidt shows three craterlets on the floor, but no traces of the small central peak which is said to stand thereon, but to be only visible in large telescopes.

SCHOMBERGER.--A large walled-plain adjoining Simpelius on the S.W. Too near the limb for satisfactory observation.

WEST LONGITUDE 20 deg. TO 0 deg.

DELAMBRE.--A conspicuous ring-plain, 32 miles in diameter, a little S. of the equator, in W. long. 17 deg. 30 min., with a massive polygonal border, terraced within, rising on the W. to the great height of 15,000 feet above the interior, but to little more than half this on the opposite side. Its outline approximates to that of a pentagon with slightly curved sides. A section on the S.E. exhibits an inflexion towards the centre. The crest is everywhere continuous except on the N., where it is broken by a deep crater with a bright rim. The north-easterly trend of the ridges and hillocks on the E. is especially noteworthy. The central peak is not prominent, but close under it on the E. is a deep fissure, extending from near the centre, and dying out before it reaches the S. border. At the foot of the N.E. glacis there are traces of a ring with low walls.

THEON, SEN.--A brilliant little ring-plain, E.N.E. of Delambre, 11 miles in diameter, and of great depth, with a regular and perfectly unbroken wall. North of it is a bright little crater.

THEON, JUN.--A ring-plain similar in size and in other respects to the last, situated about 23 miles S. of it on a somewhat dusky surface. Between the pair is a curious oblong-shaped mountain mass; and on the E. a long cliff (of no great altitude, but falling steeply on the E. side) extending S. towards Taylor a. Just below the escarpment, I find a brilliant little pair of craterlets, of which Neison only shows one.

ALFRAGANUS.--A large bright crater, about 9 miles in diameter, with very steep walls, some distance S.S.W. of Delambre, and standing on the W. edge of a large but very shallow and irregular depression W. of Taylor. There is a remarkable chain of craters on the W. of it. Alfraganus is the centre of a system of light streaks radiating in all directions, one ray extending through Cyrillus to Fracastorius.

TAYLOR.--A deep spindle-shaped ring-plain, S. of Delambre, about 22 miles in length. The wall appears to be everywhere continuous, except at the extreme N. and S. ends, where there are small craters. The outer slopes, both on the E. and W., are very broad and prominent, but apparently not terraced. There is an inconspicuous central hill. On the W. is the irregular enclosure, already referred to under Alfraganus. Three or four short winding valleys traverse the N. edge of this formation, and descend to the dark floor. On the N.E. is the remarkable ring-plain Taylor a, 18 miles in diameter, rising, at an almost isolated mountain mass on the E. border, to a height of 7000 feet above the interior. The more regular and W. section of this formation is not so lofty, and falls with a gentle slope to the dark uneven floor, on which there is some detail in the shape of small bright ridges and mounds. On the surface, N.W. of Taylor a, is a curious linear row of bright little hills. Taylor and the vicinity is better seen under low evening illumination than under morning light.

HIPPARCHUS.--Except under a low sun, this immense walled-plain is by no means so striking an object as a glance at its representation on a chart of the moon would lead one to expect; for the border, in nearly every part of it, bears unmistakable evidence of wreck and ruin, its continuity being interrupted by depressions, transverse valleys, and gaps, and it nowhere attains a great altitude. This imperfect enclosure extends 97 miles from N. to S., and about 88 miles from E. to W., and in shape approximates to that of a rhombus with curved sides. One of the most prominent bright craters on its border is Hipparchus G, on the W. Another, of about the same size, is Hipparchus E, on the N. of Horrocks. On the E. there is a moderately bright crater, Hipparchus F; and S. of this, on the same side, two others, K and I. The interior is crossed by many ridges, and near the centre includes the relics of a low ring, traversed by a narrow rill-like valley. Schmidt shows a cleft running from F across the floor to the S. border.

[A valuable monograph of Hipparchus, by Mr. W.B. Birt, was published in 1870.]

HORROCKS.--This fine ring-plain, 18 miles in diameter, stands on the N. side of the interior of Hipparchus, close to the border. It has a continuous wall, rising on the E. to a height of nearly 8000 feet above the interior, and a distinct central mountain.

HALLEY.--A ring-plain, 21 miles in diameter, on the S.W. border of Hipparchus, with a bright wall, rising at one point on the E. to a height of 7500 feet above the floor, which is depressed about 4000 feet below the surface. Two craterlets on the floor, one discovered by Birt on Rutherfurd's photogram of 1865, and the other by Gaudibert, raised a suspicion of recent lunar activity within this ring. A magnificent valley, shown in part by Schmidt as a crater-row, runs from the S. of Halley to the W. side of Albategnius.

HIND.--A ring-plain, 16 miles in diameter, a few miles W. of Halley, with a peak on its E. wall 10,000 feet above the floor. The border is broken both on the S.E. and N.E. by small craters.

[Horrocks, Halley, and Hind may be regarded as strictly belonging to Hipparchus.]

ALBATEGNIUS.--A magnificent walled-plain, 65 miles in diameter, adjoining Hipparchus on the S., surrounded by a massive complex rampart, prominently terraced, including many depressions, and crossed by several valleys. It is surmounted by very lofty peaks, one of which on the N.E. stands nearly 15,000 feet above the floor. The great ring-plain Albategnius A, 28 miles in diameter, intrudes far within the limits of the formation on the E., and its towering crest rises more than 10,000 feet above its floor, on which there is a small central mountain. The central mountain of Albategnius is more than 4000 feet high, and, with the exception of a few minor elevations, is the only prominent feature in the interior, though there are many small craters. Schmidt counted forty with the Berlin refractor, among them 12 on the E. side, arranged like a string of pearls.

PARROT.--An irregularly-shaped formation, 41 miles in diameter, S. of Albategnius, with a very discontinuous margin, interrupted on every side by gaps and depressions, large and small; the most considerable of which is the regular ring-plain Parrot a, on the E. An especially fine valley, shown by Schmidt to consist in part of large inosculating craters, cuts through the wall on the S.W., and runs on the E. side of Argelander towards Airy. The floor of Parrot is very rugged.

DESCARTES.--This object, about 30 miles in diameter, situated N.W. of Abulfeda, is bounded by ill-defined, broken, and comparatively low walls; interrupted on the S.E. by a fine crater, Descartes A, and on the S.W. by another, smaller. There is also a brilliant crater outside on the N.W. Schmidt shows a crater-row on the floor, which I have seen as a cleft.

DOLLOND.--A bright crater, about 6 miles in diameter, on the N.E. side of Descartes. Between it and the latter there is a rill-valley.

TACITUS.--A bright ring-plain, about 28 miles in diameter, a few miles E. of Catherina, with a lofty wall rising both on the E. and W. to more than 11,000 feet above the floor. Its continuity is broken on the N. by a gap occupied by a depression, and there is a conspicuous crater below the crest on the S.W. The central mountain is connected with the N. wall by a ridge, recalling the same arrangement within Madler. A range of lofty hills, an offshoot of the Altai range, extends from Tacitus towards Fermat.

ALMANON.--This ring-plain, with its companion Abulfeda on the N.E., is a very interesting telescopic object. It is about 36 miles in diameter, and is surrounded by an irregular border of polygonal shape, the greatest altitude of which is about 6000 feet above the floor on the W. It is slightly terraced, and is broken on the S. by a deep crater pertaining to the bright and large formation Tacitus b, the E. border of which casts a fine double-peaked shadow at sunrise. On the N.W. there is another bright crater, the largest of the row, running in a W.S.W. direction, and forming a W. extension of the remarkable crater-chain tangential to the borders of Almanon and Abulfeda. The only objects on the floor are three little hills, in a line, near the centre, a winding ridge on the W. side of it, and two or three other low elevations.

ABULFEDA.--A larger and more massive formation than Almanon, 39 miles in diameter, the E. wall rising about 10,000 feet above the interior, which is depressed more than 3000 feet. It is continuous on the W., but much broken by transverse valleys on the S.E., and by little depressions on the N. On the S.E. originates the very curious bright crater-row which runs in a straight line to the N.W. wall of Almanon, crossing for the first few miles the lofty table-land lying on the S.E. side of the border. With the exception of a low central mountain, the interior of Abulfeda contains no visible detail. The rampart is finely terraced on the E. and W. The E. glacis is very rugged.

ARGELANDER.--This conspicuous ring-plain, about 20 miles in diameter, is, if we except two smaller inosculating rings on the S.W. flank of Albategnius, the most northerly of a remarkable serpentine chain of seven moderately-sized formations, extending for nearly 180 miles from the S.W. of Parrot to the N. side of Blanchinus. Its border is lofty, slightly terraced within, and includes a central peak.

AIRY.--About 22 miles in diameter, connected with Argelander by a depression bounded by linear walls. Its border, double on the S.E., is broken on the S. by a prominent crater, with a smaller companion on the W. of it; and again on the N.E. by another not so conspicuous. It has a central peak.

The next link in the chain of ring-plains is Airy c, a very irregular object, somewhat larger, and with, for the most part, linear walls.

DONATI.--A ring-plain on the S. of Airy c, about 22 miles in greatest length. It is very irregular in outline, with a lofty broken border, especially on the N. and S., where there are wide gaps. There is another ring on the S.E.

FAYE.--The direction of the chain swerves considerably towards the E. at this formation, which resembles Donati both in size and in irregularity of outline. The wall, where it is not broken, is slightly terraced. There is a craterlet on the S. rim and a central crater in the interior.

DELAUNAY.--Adjoins Faye on the S.E., and is a larger and more complex object, of irregular form, with very lofty peaks on its border. A prominent ridge of great height traverses the formation from N. to S., abutting on the W. border of Lacaille. Delaunay is the last link in the chain commencing with Argelander.

LACAILLE.--An oblong enclosure situated on the N. side of Blanchinus, and apparently about 30 miles in greatest diameter. The border is to a great extent linear and continuous on the N., but elsewhere abounds in depressions. Two large inosculating ring-plains are associated with the N.E. wall.

BLANCHINUS.--A large walled-plain on the W. of Purbach and abutting on the S. side of Lacaille. It much resembles Purbach in shape, but has lower walls. Schmidt shows a crater on the N. side of the floor, which I have seen, and a number of parallel ridges which have not been noted, probably because they are only visible under very oblique light.

GEBER.--A bright ring-plain, 25 miles in diameter, S. of Almanon, with a regular border, rising to a height on the W. of nearly 9000 feet above the floor. There is a small crater on the crest of the S. wall, and another on the N. A ring-plain about 8 miles in diameter adjoins the formation on the N.E. According to Neison, there is a feeble central hill, which, however, is not shown by Schmidt.

SACROBOSCO.--This is one of those extremely abnormal formations which are almost peculiar to certain regions in the fourth quadrant. It is about 50 miles in greatest diameter, and is enclosed by a rampart of unequal height, rising on the E. to 12,000 feet above the floor, but sinking in places to a very moderate altitude. On the N. its contour is, if possible, rendered still more irregular by the intrusion of a smaller ring-plain. On the N.E. side of the floor stands a very bright little crater and two others on the S. of the centre, each with central mountains.

FERMAT.--An irregular ring-plain 25 miles in diameter on the W. of Sacrobosco. Its partially terraced wall is broken on the N. by a gap which communicates with the interior of a smaller formation. There are some low hills on the floor, which is depressed 6000 feet below the crest of the border.

AZOPHI.--A prominent ring-plain, 30 miles in diameter, E.N.E. of Sacrobosco, its lofty barrier towering nearly 11,000 feet above a somewhat dusky interior, which includes some light spots. A massive curved mountain arm runs from the S. side of this formation to a small ring-plain W. of Playfair.

ABENEZRA.--When observed near the morning terminator, this noteworthy ring-plain, 27 miles in diameter, seems to be divided into two by a curved ridge which traverses the formation from N. to S., and extends beyond its limits. The irregular border rises on the W. to a height of more than 14,000 feet above the deeply-sunken floor, which includes several craters, hills, and ridges.

APIANUS.--A magnificent ring-plain, 38 miles in diameter, N.W. of Aliacensis, with lofty terraced walls, rising on the N.E. to about 9000 feet above the interior, and crowned on the W. by three large conspicuous craters. The border is broken on the N. by a smaller depression and a large ring with low walls. The dark-grey floor appears to be devoid of conspicuous detail.

PLAYFAIR.--A ring-plain, 28 miles in diameter, with massive walls. It is situated on the N. of Apianus, and is connected with it by a mountain arm. The rampart is tolerably continuous, but varies considerably in altitude, rising on the S. to a height of more than 8000 feet above the interior. On the E., extending towards Blanchinus, is a magnificent unnamed formation, bounded on the E. by a broad lofty rampart flanking Blanchinus, Lacaille, Delaunay, and Faye; and on the W. by Playfair and the mountain arm just mentioned. It is fully 60 miles in length from N. to S. Sunrise on this region affords a fine spectacle to the observer with a large telescope. The best phase is when the morning terminator intersects Aliacensis, as at this time the long jagged shadows of the E. wall of Playfair and of the mountain arm are very prominent on the smooth, greyish-blue surface of this immense enclosure.

PONTANUS.--An irregular ring-plain, 28 miles in diameter, S.S.W. of Azophi, with a low broken border, interrupted on the S.W. by a smaller ring-plain, which forms one of a group extending towards the S.W. The dark floor includes a central mountain.

ALIACENSIS.--This ring-plain, 53 miles in diameter, with its neighbour Werner on the N.E., are beautiful telescopic objects under a low sun. Its lofty terraced border rises at one peak on the E. to the tremendous height of 16,500 feet, and at another on the opposite side to nearly 12,000 feet above the floor. The wall on the S. is broken by a crater, and on the W. traversed by narrow passes. There is also a prominent crater on the inner slope of the N.E. wall. The floor includes a small mountain, several little hills, and a crater.

WERNER.--A ring-plain, 45 miles in diameter, with a massive rampart crowned by peaks almost as lofty as any on that of Aliacensis, and with terraces fully as conspicuous. It has a magnificent central mountain, 4500 feet high. At the foot of the N.E. wall Madler observed a small area, which he describes as rivalling the central peak of Aristarchus in brilliancy. Webb, however, was unable to confirm this estimate, though he noted it as very bright, and saw a minute black pit and narrow ravine within it. Neison subsequently found that the black pit is a crater-cone. It would perhaps be rash, with our limited knowledge of minute lunar detail, to assert that Madler over-estimated the brightness of this area, which may have been due to a recent deposit round the orifice of the crater-cone.

POISSON.--An irregular formation on the W. of Aliacensis, extending about 50 miles from W. to E., but much less in a meridional direction. Its N. limits are marked by a number of overlapping ring-plains and craters, and it is much broken elsewhere by smaller depressions. The E. wall is about 7000 feet in height.

GEMMA FRISIUS.--A great composite walled-plain, 80 miles or more in length from N. to S., with a wall rising at one place nearly 14,000 feet above the floor. It is broken on the N. by two fine ring-plains, each about 20 miles in diameter, and on the E. by a third open to the E. There is a central mountain, and several small craters on the floor, especially on the W. side.

BUSCHING.--A ring-plain S. of Zagut, about 36 miles in diameter, with a moderately high but irregular wall. There are several craterlets within and some low hills.

BUCH.--Adjoins Busching on the S.E. It is about 31 miles in diameter, and has a less broken barrier. There is a large crater on the E. wall, and another smaller one on the S.W. Schmidt shows nothing on the floor, but Neison noted two minute crater-cones.

MAUROLYCUS.--This unquestionably ranks as one of the grandest walled-plains on the moon's visible surface, and when viewed under a low sun presents a spectacle which is not easily effaced from the mind. Like so many of the great enclosures in the fourth quadrant, it impresses one with the notion that we have here the result of the crowding together of a number of large rings which, when they were in a semi-fluid or viscous condition, mutually deformed each other. It extends fully 150 miles from E. to W., and more from N. to S.; so it may be taken to include an area on the lunar globe which is, roughly speaking, equal to half the superficies of Ireland. This vast space, bounded by one of the loftiest, most massive, and prominently-terraced ramparts, includes ring-plains, craters, crater-rows, and valleys,--in short, almost every type of lunar formation. It towers on the E. to a height of nearly 14,000 feet above the interior, and on the W., according to Schmidt, to a still greater altitude. A fine rill-valley curves round the outer slope of the W. wall, just below its crest, which is an easy object in a 8 1/2 inch reflector when the opposite border is on the morning terminator, and could doubtless be seen in a smaller instrument; and there is an especially brilliant crater on the S. border, which is not visible till a somewhat later stage of sunrise. The central mountain is of great altitude, its loftiest peaks standing out amid the shadow long before a ray of sunlight has reached the lower slopes of the walls. It is associated with a number of smaller elevations. I have seen three considerable craters and several smaller ones in the interior.

BAROCIUS.--A massive formation, about 50 miles in diameter, on the S.W.

side of Maurolycus, whose border it overlaps and considerably deforms. Its wall rises on the E. to a height of 12,000 feet above the floor, and is broken on the N.W. by two great ring-plains. On the inner slope of the S.E. border is a curious oblong enclosure. There is nothing remarkable in the interior. On the dusky grey plain W. of Maurolycus and Barocius there is a number of little formations, many of them being of a very abnormal shape, which are well worthy of examination. I have seen two short unrecorded clefts in connection with these objects.

STOFLER.--A grand object, very similar in size and general character to Maurolycus, its neighbour on the W. To view it and its surroundings at the most striking phase, it should be observed when the morning terminator lies a little E. of the W. wall. At this time the jagged, clean-cut, shadows of the peaks on Faraday and the W. border, the fine terraces, depressions, and other features on the illuminated section of the gigantic rampart, and the smooth bluish-grey floor, combine to make a most beautiful telescopic picture. At a peak on the N.E., the wall attains a height of nearly 12,000 feet, but sinks to a little more than a third of this height on the E. It is apparently loftiest on the N. The most conspicuous of the many craters upon it is the bright deep circular depression E. on the S. wall, and another, rather larger and less regular, on the N.W., which has a very low rim on the side facing the floor, and a craterlet on either side of the apparent gap. A large lozenge-shaped enclosure abuts on the wall, near the crater E., with a border crowned by a number of little peaks, which at an early stage of sunrise resemble a chaplet of pearls. The floor of Stofler is apparently very level, and in colour recalls the beautiful steel-grey tone of Plato seen under certain conditions. I have noted several distinct little craters on its surface, mostly on the N.E. side; and on the E. side a triangular dark patch, close to the foot of the wall, very similar in size and appearance to those within Alphonsus.

FARADAY.--A large ring-plain, about 35 miles in diameter, overlapping the S.W. border of Stofler; its own rampart being overlapped in its turn by two smaller ring-plains on the S.E., and by two still smaller formations (one of which is square-shaped) on the N.W. The wall is broad and very massive on the E. and N.E., prominently terraced, and includes many brilliant little craters. Schmidt shows a ridge and several craters in the interior.

LICETUS.--An irregular formation, about 50 miles in maximum width, on the

S. of Stofler, with the flanks of which it is connected by a coarse valley. Neison points out that it consists of a group of ring-plains united into one, owing to the separating walls having been partially destroyed. This seems to be clearly the case, if Licetus is examined under a low sun. On the E. side of the N. portion of the formation, the wall rises to nearly 13,000 feet.

FERNELIUS.--A ring-plain, about 30 miles in diameter, abutting on the N. wall of Stofler. It is overlapped on the E. by another similar formation of about half its size. There are many craters and depressions on the borders of both, and a large crater between the smaller enclosure and the N.E. outer slope of Stofler. Schmidt shows eight craters on the floor of Fernelius.

NONIUS.--A ring-plain, about 20 miles in diameter, abutting on the N. wall of Fernelius. There is a prominent bright crater on the W. of it, and another on the N., from which a delicate valley runs towards the W. side of Walter.

CLAIRAUT.--A very peculiar formation, about 40 miles in diameter, S. of Maurolycus, affording another good example of interference and overlapping. The continuity of its border, nowhere very regular, has been entirely destroyed on the S. by the subsequent formation of two large rings, some 10 or 12 miles in diameter, the more easterly of which has, in its turn, been partially wrecked on the N. by a smaller object of the same class. There is also a ring-plain N.E. of Clairaut, which has very clearly modified the shape of the border on this side. Two craters on the floor of Clairaut are easy objects.

BACON.--A very fine ring-plain, 40 miles in diameter, S.W. of Clairaut. At one peak on the E. the terraced wall rises to nearly 14,000 feet above the interior. It is broken on the S. by three or four craters. On the W. there is an irregular inconspicuous enclosure, whose contiguity has apparently modified the shape of the border. There are two large rings on the N. (the more easterly having a central peak), and a third on the E. The floor appears to be devoid of prominent detail.

CUVIER.--A walled-plain, about 50 miles in diameter, on the S.E. of Clairaut. The border on the E. rises to 12,000 feet; and on the N.W. is much broken by depressions. Neison has seen a mound, with a minute crater W. of it, on the otherwise undisturbed interior.

JACOBI.--A ring-plain S. of Cuvier, about 40 miles in diameter, with walls much broken on the N. and S., but rising on the E. to nearly 10,000 feet. There is a group of craters (nearly central) on the floor. The region S. of this formation abounds in large unnamed objects.

LILIUS.--An irregular ring-plain, 39 miles in diameter, with a rampart on the E. nearly 10,000 feet above the floor. A smaller ring between it and Jacobi has considerably inflected the wall towards the interior. It has a conspicuous central mountain.

ZACH.--A massive formation, 46 miles in diameter, on the S. of Lilius, with prominently terraced walls, rising on the E. to 13,000 feet above the interior. A small ring-plain, whose wall stands 6000 feet above the floor, is associated with the N. border. Two other rings, on the S.W. and N.E. respectively, have craters on their ramparts and central hills.

PENTLAND.--A fine conspicuous formation under a low sun, even in a region abounding in such objects. It is about 50 miles in diameter, with a border exceeding in places 10,000 feet in height above the floor, which includes an especially fine central mountain.

KINAU.--One of the group of remarkable ring-plains extending in a N.W. direction from Pentland.

SIMPELIUS.--Another grand circumvallation, almost as large as Pentland, but unfortunately much foreshortened. One of its peaks on the E. rises to a height of more than 12,000 feet above the floor, on which there is a small central mountain. Between Simpelius and Pentland are several ring- plains, most of which appear to have been squeezed and deformed into abnormal shapes.

CURTIUS.--A magnificent formation, about 50 miles in diameter, with one of the loftiest ramparts on the visible surface, rising at a mountain mass on the N.E. to more than 22,000 feet, an altitude which is only surpassed by peaks on the walls of Newton and Casatus. There is a bright crater on the S.E. border and another on the W. The formation is too near the S. limb for satisfactory scrutiny. Between Curtius and Zach is a fine group of unnamed enclosures.

APPENDIX

DESCRIPTION OF THE MAP

The accompanying map, eighteen inches in diameter, represents the moon under mean libration. Meridian lines and parallels of latitude are drawn at every 10 deg., except in the case of the meridians of 80 deg. E. and W. longitude, which are omitted to avoid confusion, and as being practically needless. These lines will enable the observer, with the aid of the Tables in the Appendix, to find the position of the terminator at any time required. As astronomical telescopes exhibit objects inverted, maps of the moon are always drawn upside down, and with the right and left interchanged, as in the diagram above, which also shows how the quadrants are numbered.

This circle [drawing of circle], intended to be .15708 in diameter, represents a circle of one degree in diameter at the centre of the map, and as the length of one selenographical degree is 18.871 miles, it represents an area of nearly 280 square miles.

The catalogue is so arranged that, beginning with the W. limb, and referring to the lists under the first and fourth, and the second and third quadrants, all the formations falling within the meridians 90 deg. to 60 deg., 60 deg. to 40 deg., 40 deg. to 20 deg., 20 deg. to 0 deg. (the central meridian), and from 0 deg. to 20 deg., and so on, to the E. limb, will be found in convenient proximity in the text.

In the Catalogue, N. S. E. W. are used as abbreviations for the cardinal points.

LIST OF THE MARIA, OR GREY PLAINS, TERMED "SEAS," &c.

FIRST QUADRANT.

Mare Tranquilitatis (nearly the whole), page 5. ,, Foecunditatis (the N. portion), 5. ,, Serenitatis, 5. ,, Crisium, 6. ,, Frigoris (a portion), 5. ,, Vaporum (nearly the whole), 6. ,, Humboldtianum, 6. ,, Smythii (a portion), 39. Lacus Mortis, 53. ,, Somniorum. Palus Somnii. ,, Nebularum (a portion), 62. ,, Putredinis, 61. Sinus Medii (a portion), 6.

SECOND QUADRANT.

Mare Imbrium, 5. ,, Nubium (the N. portion), 5. ,, Frigoris (a portion), 5. ,, Vaporum (a portion), 6. Oceanus Procellarum (the N. portion), 5. Palus Nebularum (a portion), 62. Sinus Iridum, 80. ,, Medii (a portion), 6. ,, Roris, 90. ,, Aestuum.

THIRD QUADRANT.

Mare Nubium (the greater portion), 5. ,, Humorum, 6. Oceanus Procellarum (the S. portion), 5. Sinus Medii (a small portion), 6.

FOURTH QUADRANT.

Mare Foecunditatis (the greater portion), 5. ,, Nectaris, 7. ,, Tranquilitatis (a small portion), 5. ,, Australe, 127. ,, Smythii (a portion), 39. Sinus Medii (a portion), 6.

LIST OF SOME OF THE MOST PROMINENT MOUNTAIN RANGES, PROMONTORIES, ISOLATED MOUNTAINS, AND REMARKABLE HILLS.

FIRST QUADRANT.

The Alps. The western portion of the range.

The Apennines. The extreme northern part of the range.

The Caucasus.

The Haemus.

The Taurus.

The North Polar Range. On the limb extending from N. lat. 81 deg. towards the E.

The Humboldt Mountains. On the limb from N. lat. 72 deg. to N. lat. 53 deg.

Mount Argaeus. A mountain mass rising some 8000 feet above the Mare Serenitatis in N. lat. 20 deg., W. long. 28 deg., N.W. of Dawes.

Prom. Acherusia. A bright promontory at the W. extremity of the Haemus range, rising nearly 5000 feet above the Mare Serenitatis. N. lat. 17 deg., W. long. 22 deg.

Cape Agarum. The N. end of a projecting headland on the S.W. side of the Mare Crisium, in N. lat. 14 deg., W. long. 66 deg., rising nearly 11,000 feet above the Mare.

Le Monnier A. An isolated mountain more than 3000 feet high, standing about midway between the extremities of the bay: probably a relic of a once complete ring.

Secchi. South of this formation there is a lofty prominent isolated mountain.

Manilius A and beta. Two conspicuous mountains N. of Manilius; A, the more westerly, being more than 5000 feet, and beta about 2000 feet in height.

Autolycus A. A mountain of considerable altitude, S. of this formation.

Mont Blanc. Principal peak, N. lat. 46 deg., W. long. 0 deg. 30 min., nearly 12,000 feet in height.

Cassini epsilon and delta. Two adjoining mountain masses N. of Cassini, more than 5000 feet high.

Eudoxus. S.E. of this formation, in N. lat. 43 deg., W. long. 10 deg., are two bright mountain masses, the more southerly rising 7000, and the other 4000 feet above the surface.

Mount Hadley. The northern extremity of the Apennines, in N. lat. 27 deg. W. long. 5 deg., rising more than 15,000 feet above the Mare.

Mount Bradley. A promontory of the Apennines, in N, lat. 23 deg., W. long. 1

deg., nearly 14,000 feet above the Mare Imbrium.

The Silberschlag Range, running from near the S.E. side of Julius Caesar to the region W. of Agrippa.

SECOND QUADRANT.

The Alps. The eastern and greater portion.

The Apennines. Nearly the whole of the range.

The Carpathians.

The Teneriffe Mountains. S.E. of Plato. Highest peak, 8000 feet.

The Straight Range. East of the last, in N. lat. 48 deg., E. long. 20 deg.

The Harbinger Mountains. N.W. of Aristarchus.

The Hercynian Mountains. Near the N.E. limb, E. of Otto Struve, N. lat. 25 deg.

Mount Huygens. A mountain mass projecting from the escarpment of the Apennines, in N. lat. 20 deg., E. long. 3 deg., one peak rising to 18,000 feet above the Mare Imbrium.

Mount Wolf. A great square-shaped mountain mass, near the S.E. extremity of the Apennines, in N. lat. 17 deg., E. long. 9 deg., the loftiest peak rising to nearly 12,000 feet above the Mare Imbrium.

Eratosthenes I and X. Two isolated mountains N. of this formation, in N. lat. 20 deg.; X is 1800 feet in height.

Pico. A magnificent isolated mountain, S. of Plato, in N. lat. 45 deg., E. long. 9 deg., rising some 8000 feet above the Mare Imbrium.

Pico B. A triple-peaked mountain a few miles S. of Pico.

Piton. A bright isolated mountain 7000 feet high, in N. lat. 1 deg., E. long. 1 deg.

Fontinelle A. A conspicuous isolated mountain about 3000 feet high, S. of Fontinelle.

Archimedes Z. A triangular-shaped group E. of Archimedes, in N. lat. 31 deg., E. long. 8 deg., the highest of the peaks rising more than 2000 feet.

Caroline Herschel. E. of this formation is a double-peaked mountain rising to 1300 feet.

Gruithuisen delta and gamma. On the N. of this bright crater, in N. lat. 36 deg., E. long. 40 deg., rises a fine mountain, delta, nearly 6000 feet in height, and on the N.E. of it the larger mass gamma, almost as lofty.

Mairan. There is a group of three bright little mountains, the loftiest about 800 feet above the Mare, some distance E. of this formation.

Euler beta. A fine but small mountain group, more than 3600 feet high, on the Mare Imbrium, S.E. of Euler.

The Laplace Promontory. A magnificent headland on the N. side of the Sinus Iridum, rising about 9000 feet above the latter, and about 7000 feet above the Mare Imbrium.

Cape Heraclides. A fine but less prominent headland on the opposite side of the bay, rising more than 4000 feet above it.

Lahire. A large bright isolated mountain in the Mare Imbrium, N.E. of Lambert, in N. lat. 27 deg., E. long. 25 deg. It is, according to Schroter, nearly 5000 feet high.

Delisle beta. A curious club-shaped mountain on the S.E. of this formation, nearly 4000 feet in height.

Pytheas beta. An isolated mountain, 900 feet high, in N. lat. 20 deg., E. long. 23 deg.

Kirch. There is a small isolated hill a few miles N. of this formation.

Kirch GAMMA. A bright mountain about 700 feet high, in N. lat. 39 deg., E. long. 3 deg.

Piazzi Smyth beta. A small bright isolated mountain on a ridge S. of this, is a noteworthy object under a low sun.

Lambert GAMMA. In N. lat. 26 deg., E. long. 18 deg.; a remarkable curved mountain about 3000 feet in height, a brilliant object under a low sun.

D'Alembert Mountains. A range on the E. limb running S. from N. lat. 12 deg.

Wollaston. An isolated triangular mountain about midway between this and Wollaston B.

THIRD QUADRANT.

The Riphaean Mountains. An isolated range S. of Landsberg in S. lat. 7 deg., E. long. 28 deg. They run in a meridional direction, and rise at one peak to nearly 3000 feet above the Oceanus Procellarum.

The Percy Mountains extend from the eastern flank of Gassendi towards Mersenius, forming the north-eastern border of the Mare Humorum.

Prom. Aenarium. A steep bluff situated at the northern end of a plateau, some distance E. of Arzachel, in S. lat. 18 deg., E. long. 9 deg. It rises some 2000 feet above the Mare Nubium.

Euclides zeta and chi. Two mountain masses N. of this formation in S. lat. 5 deg.; zeta rises about 1700 feet above the Mare; both are evidently offshoots from the Riphaean range.

Landsberg H. An isolated hill in S. lat. 4 deg., E. long. 25 deg.

Nicollet C. S.E. of Nicollet, in S. lat. 22 deg., E. long. 17 deg.; is hemmed in by a mountain mass rising to more than 2000 feet above the Mare Nubium.

The Stag's-Horn Mountains. At the S. end of the straight wall, or "railroad," in S. lat. 24 deg., E. long. 8 deg., a curious mountain mass rising about 2000 feet above the Mare Nubium.

Lacroix delta. A mountain more than 7000 feet high, N. of Lacroix.

Flamsteed E. A mountain of more than 3000 feet in S. lat. 4 deg., E. long. 51 deg.

D'Alembert Mountains. A very lofty range on the E. limb, extending to S. lat. 11 deg.

The Cordilleras. Close to the E. limb; they lie between S. lat. 8 deg. and S. lat. 23 deg.

Rook Mountains. On the E. limb, extending from about S. lat. 18 deg. to S. lat. 35 deg. According to Schroter, they attain a height of 25,000 feet.

Dorfel Mountains. On the S.E. limb between S. lat. 57 deg. and S. lat. 80 deg.

Leibnitz Mountains. On the S. limb extending W. from S. lat. 80 deg. beyond the Pole on to the Fourth Quadrant. Perhaps the loftiest range on the limb. Madler's measures give more than 27,000 feet as the height of one peak, and there are several others nearly as high.

FOURTH QUADRANT.

The Altai Mountains. A fine conspicuous serpentine range, extending from the E. side of Piccolomini in a north-easterly direction to the region between Tacitus and Catherina, a length of about 275 miles. The loftiest peak is over 13,000 feet. The average height of the southern portion is about 6000 feet. The region lying on the S.E. of this range is a vast tableland, devoid of prominent objects, rising gradually towards the mountains, which shelve rapidly down to an equally barren expanse on the N.W.

The Pyrenees. These mountains, on the E. of Guttemberg, border the western side of the Mare Nectaris. Their loftiest peak, rising nearly to 12,000 feet, is on the S.E. of Guttemberg.

LIST OF THE PRINCIPAL RAY-SYSTEMS, LIGHT-SURROUNDED CRATERS, AND LIGHT- SPOTS.

[In this list, which does claim to be exhaustive, most of the objects noted by Schmidt are incorporated.]

FIRST QUADRANT.

Autolycus. Encircled by a delicate nimbus, throwing out four or five prominent rays extending towards Archimedes. Seen best under evening illumination.

Aristillus. The centre of a noteworthy system of delicate rays extending W. towards the Caucasus; and on the S. disappearing among the rays of Autolycus. They are traceable on the Mare Nubium near Kirch.

Theaetetus. A very brilliant group of little hills E. of this formation.

Eudoxus A. A light-surrounded crater W. of Eudoxus, with distinct long streaks, one of which extends to the S. wall of Aristoteles.

Aristoteles A. A light-surrounded crater in the Mare Frigoris, N.E. of Aristoteles.

Aratus. A very conspicuously brilliant crater in the Apennines, with a smaller light-surrounded crater W. of it.

Sulpicius Gallus. A light spot near.

Manilius. Surrounded by a light halo and streaks.

Taquet. Has a prominent nimbus, and indications of very delicate streaks.

Plinius A. Is surrounded by a well-marked halo.

Posidonius gamma. Among the hills E. of this formation a light spot resembling Linne, according to Schmidt. He first saw it in 1867, when it had a delicate black spot in the centre. Dr. Vogel observed and drew it in 1871 with the great refractor at Bothkamp. These observations were confirmed by Schmidt in 1875 with the 14-feet refractor at Berlin.

Littrow. A very bright light-spot with streaks, on the site of a little crater and well-known cleft E. of this ring-plain.

Romer. A light-surrounded mountain on the E.

Macrobius. Two light-surrounded craters on the E. of this formation, the more northerly being the brighter.

Cleomedes A. (On the floor.) Surrounded by a nimbus and rays. Large crater, A, on the E. has also a nimbus and rays.

Agrippa. Exhibits faint rays.

Godin. Exhibits faint rays.

Proclus. A well-known ray-centre, some of the rays prominent on part of the Mare Crisium.

Taruntius. Has a very faint nimbus, with rays, on a dark surface.

Dionysius. A brilliant crater with a prominent, bright, excentrically placed nimbus on a dark surface, on which distinct rays are displayed.

Hypatia B. A very small bright crater on a dark surface: surrounded by a faint nimbus.

Apollonius. Among the hills S. of this, there is a small bright streak system.

Eimmart. There is a large white spot N.W. of this.

Geminus is associated with a system of very delicate rays.

Menelaus. A brilliant object. It is traversed by a long ray from Tycho.

SECOND QUADRANT.

Anaxagoras. The centre of an important ray-system.

Timocharis is surrounded by a pale irregular nimbus and faint rays, most prominently developed on the W. side of the formation.

Copernicus. Next to Tycho, the most extended ray-centre on the visible surface. Some distance on the E., in E. long. 25 deg., N. lat. 11 deg., lies a very small but conspicuous system, and in E. long. 22 deg., N. lat. 8 deg. a bright light spot among little hills.

Gambart A. A bright crater with large nimbus and rays.

Landsberg A. A light-surrounded crater on a dark surface, with companions, referred to under the Third Quadrant.

Encke. There is a light-surrounded crater S. of this.

Kepler. A noted ray-centre. It is surrounded by an extensive halo, especially well developed on the E., across the Mare Procellarum.

Bessarion. Two bright craters: the more northerly is prominently light-surrounded, while its companion is less conspicuously so.

Aristarchus.--The most conspicuous bright centre on the moon, the origin of a complicated ray-system.

Delisle. S. of this formation there is a tolerably bright spot on the site of some hills.

Timaeus. A ray-centre.

Euler. Feeble halo with streaks.

Galileo. Between this and Reiner is a curious bright formation with short rays, referred to in the Catalogue, under Reiner.

Cavalerius. A light streak originating in the W. wall, and extending on to the Oceanus Procellarum.

Olbers. A considerable ray-system, but seldom distinctly visible.

Lichtenberg. Faintly light-surrounded.

THIRD QUADRANT.

Tycho. The largest and best known system on the visible surface.

Zuchius. A remarkable ray-system, but one which is only well seen when libration is favourable.

Bailly. N. of the centre of this great enclosure are two very distinct radiating streaks.

Schickard. Four conspicuous light spots, probably craters, on the S.E.

Byrgius A. A brilliant ray-centre, most of the rays trending eastward from a nimbus.

Hainzel. There are several bright spots E. of this formation.

Mersenius. Two or three light-rays originate from a point on the W. rampart.

Mersenius C. A light-surrounded crater with short rays.

Grimaldi. There are three bright spots on the W. wall.

Damoiseau. A light-surrounded crater W. of Damoiseau, E. long. 58 deg., S. lat. 6 deg.

Flamsteed C. A light-surrounded crater on a dark surface.

Lubieniezky A. Crater with halo on a dark surface.

Lubieniezky F. Crater with halo on a dark surface.

Lubieniezky G. Crater with halo on a dark surface.

Birt a. A light-surrounded crater.

Landsberg. E. of Landsberg, four light-surrounded craters, forming with Landsberg A (in the Second Quadrant) an interesting group.

Lohrmann A. A light-surrounded crater, with a light area a few miles N. of it. S. lat. 1 deg., E. long. 61 deg.

Euclides. Has a conspicuous nimbus with traces of rays, a typical example.

Guerike. There is a crater, with nimbus, W. of this, in E. long. 12 deg., S. lat. 11 deg. 5 min.

Parry. A very brilliant light-spot in the S. wall.

Parry A. Surrounded by a bright nimbus.

Alpetragius B. A conspicuous light-surrounded crater, one of the most remarkable on the moon.

Alpetragius d (E. long. 11 deg., S. lat. 13 deg. 8 min.). A bright spot, seen by Madler as a crater, but which, as Schmidt found in 1868, no longer answers to this description.

Mosting C. A light-surrounded crater.

Lalande. Has a large nimbus and distinct rays.

Hell. A large ill-defined spot in E. long. 4 deg., S. lat. 33 deg. This is most probably the site of the white cloud seen by Cassini.

Mercator. There is a brilliant crater and light area under E. wall.

FOURTH QUADRANT.

Stevinus a. A crater E. of Stevinus; it is a centre of wide extending rays.

Furnerius A. Prominently light-surrounded, with bright streaks, radiating for a long distance N. and S.

Messier A. The well-known "Comet" rays, extending E. of this.

Langrenus. Has a large but very pale ray-system. It is best seen under a low evening sun. Three long streaks radiate towards the E. from the foot of the glacis of the S.E. wall.

Censorinus. A very brilliant crater with faint rays.

Theophilus. The central mountain is faintly light-surrounded.

Madler. This ring-plain and the neighbourhood on the N. and N.W., include many bright areas and curious streaks.

Almanon. About midway between this and Argelander is a very brilliant little crater.

Beaumont. Between this and Cyrillus stand three considerable craters with nimbi.

Cyrillus A. A prominent light-surrounded crater.

Alfraganus. A light-surrounded crater with rays.

POSITION OF THE LUNAR TERMINATOR

Though the position of the Lunar Terminator is given for mean midnight throughout the year in that very useful publication the Companion to the Observatory, it is frequently important in examining or comparing former drawings and observations to ascertain its position at the times when they were made. For this purpose the subjoined tables (which first appeared in the Selenographical Journal) will be found useful, as they give for any day between A.D. 1780 and A.D. 1900 the selenographical longitude of the point where the terminator crosses the moon's equator, which it does very nearly at right angles.

[Tables and examples]

LUNAR ELEMENTS

Moon's mean apparent diameter - 31 min. 8 sec.

Moon's maximum apparent diameter - 33 min. 33.20 sec.

Moon's minimum apparent diameter - 29 min. 23.65 sec.

Moon's diameter, in miles - 2163 miles.

Volume (earth's = 1) - 1/49.20 or 0.02033.

Mass (earth's = 1) - 1/81.40 or 0.0128.

Density (earth's = 1) - 0.60419, or 3.444 the density of water (water being unity).

Surface area, about 14,600,000 square miles (earth's surface area, 196,870,000 miles)

Earth's surface area = 1, moon's - About 2/27 or 0.07407.

Action of gravity at surface - 0.16489 or 1/6.065 of the earth's.

Surface of moon never seen - 0.4100.

Surface of moon seen at one time or another - 0.5900.

Synodical revolution, or interval from new moon to new moon (commonly called a lunation) - 29 d. 12 h. 44 m. 2.684 s. - 29.5305887 days.

Sidereal revolution, or time taken in passing from one star to the same star again - 27 d. 7 h. 43 m. 11.545 s. - 27.3216614 days.

Tropical revolution, or time taken in passing from "the first point of Aries" to the same point again - 27 d. 7 h. 43 m. 4.68 s. - 27.321582 days.

Anomalistic revolution, or time taken in passing from perigee to perigee - 27 d. 13 h. 18 m. 37.44 s. - 27.55460 days.

Nodical revolution, or time taken in passing from rising node to rising node - 27 d. 5h. 5m. 35.81 s. - 27.21222 days.

Distance (mean) in terms of the equatorial radius of the earth - 60.27.

Distance in miles (mean) - 238,840 miles.

Distance, maximum - 252,972 miles.

Distance, minimum - 221,614 miles.

Mean excentricity of moon's orbit - 0.05490807.

Inclination of moon's orbit to the ecliptic (mean) - 5 deg. 8 min. 39.96 sec.

Inclination of moon's axis to the ecliptic - 87 deg. 27 min. 51 sec.

Inclination of moon's equator to the ecliptic - 1 deg. 32 min. 9 sec.

Maximum libration in latitude - 6 deg. 44 min.

Maximum libration in longitude - 7 deg. 45 min.

Maximum total libration from earth's centre - 10 deg. 16 min.

Maximum diurnal libration - 1 deg. 1 min. 28.8 sec.

Angle subtended by one degree of selenographical latitude and longitude at the centre of the moon's disc, when at its mean distance - 16.566 sec.

Length of a degree under these conditions - 18.871 miles.

Selenographical arc at the centre of the moon's surface, subtending an angle of one second of arc - 3 min. 37.31 sec.

Miles at the centre of the moon's disc, subtending an angle of one second of arc - 1.139

[It must be remembered that this value is increased, in departing from the centre, in the proportion of the secants of the angular distance from the centre.]

Period of similar phase - 59 d. 1h. 28m. = 2 lunations.

Or, more accurately - 442 d. 23 h. = 15 lunations.

###

www.ingramcontent.com/pod-product-compliance
Lightning Source LLC
Chambersburg PA
CBHW070856180526
45168CB00005B/1847